Green Gentrification and Environmental Injustice

Heather E. Campbell • Adam Eckerd
Yushim Kim

Green Gentrification and Environmental Injustice

A Complexity Approach to Policy

Springer

Heather E. Campbell
Claremont Graduate University
Claremont, CA, USA

Adam Eckerd
Old Dominion University
Norfolk, VA, USA

Yushim Kim
Arizona State University
Phoenix, AZ, USA

ISBN 978-3-031-65099-4 ISBN 978-3-031-65100-7 (eBook)
https://doi.org/10.1007/978-3-031-65100-7

This Springer imprint is published by the registered company Springer Nature Switzerland AG
The registered company address is: Gewerbestrasse 11, 6330 Cham, Switzerland

Acknowledgments

We are very grateful for the excellent research assistance provided by Jessica Geiger and Jacqueline "Yvette" Hardy.

Heather E. Campbell also appreciates the sabbatical support provided by Claremont Graduate University, and especially by Professor Melissa Rogers, who took over Heather's administrative duties so that she could work on this book. She is also appreciative of support from Tiffany Berry, Dean of the School of Social Science, Policy & Evaluation, and Division staff Lidia Romero-Arriaza and Cara Wong. Heather also thanks her parents for assistance with the index. Yushim Kim appreciates the sabbatical support provided by Arizona State University.

Contents

1 Introduction to Green Gentrification and Our Approach 1
Introduction 1
Gentrification and Displacement 2
Recent Gentrification and Displacement in the US 3
A Gentrification Framework: Government Intervention 4
Can There Be Gentrification Without Displacement? 6
Potential Causes of Gentrification 8
Mitigators of Gentrifier Inflow 10
Mitigators of Displacee Outflow 10
Policy Actions 11
Street-Level Insights 13
Discussions and Conclusion 14
References 16

2 The Environmental Injustice Problem and Its Links to Gentrification and Displacement: Literature Review 19
Introduction 19
A Brief Overview of EJ Literature 23
EJ and Health 24
Brief Review of Gentrification and Displacement 26
Overview of Green Gentrification Literature 30
Mixed Evidence of Green Displacement 31
Proposed Solutions 33
Just Green Enough 34
What Are We Doing Differently? 36
References 38

3 Complexity Approach to Urban Systems 43
Introduction 43
Complex Adaptive Systems View 46
Neighborhood Change 46
CAS Research Framework 48

Two Research Approaches for the Urban CAS 51
Qualitative Comparative Analysis (QCA) 52
Agent-Based Modeling (ABM) 54
Conclusion 56
References 57

4 **Conceptual Clarification: Gentrification, and Displacement** 61
Introduction 61
Gentrification as a Complex Process 62
Gentrification Phases 64
Gentrification-Led Displacement 65
Interacting Processes 67
Policy Implications 68
Policy Vantage Points 70
Conclusion 71
References 71

5 **New Amenities that Provide Jobs: Consequences of Density and Segregation** 73
Introduction 73
Gentrification 74
The Agent-Based Model 78
Model Structure 79
Residents 80
Land Uses 83
Land 84
Defining Model Outcomes 85
Scenarios and Setup 86
Results 88
Policy Choices 88
Residential Preferences 88
Discussion and Policy Implications 93
Policy Implications 94
Conclusion 97
Appendix: Model Robustness Assessments 97
References 100

6 **Clusters of Conditions for Green Gentrification** 105
Jieun Kim and Michelle Stuhlmacher
Introduction 105
Green Gentrification Research 106
Qualitative Comparative Analysis (QCA) Approach to Green Gentrification 109
QCA as a Method 111
A Case Study of Chicago Neighborhoods in the 2010s 113
Case Selection 113

Outcome Condition: Gentrification (Gent) 114
Explanatory Conditions 115
Findings 117
Path 1 "Green Growth Machine": LPARK*igreen*GHOUS → GENT 119
Path 2 "Small-Park-Involved Gentrification": SPARK*igreen*ghous → GENT 119
Path 3 "Spillover Gentrification": NOPARK*igreen*GHOUS *GNEIGH → GENT 119
Discussion 120
Conclusion 122
Appendixes 123
References 127

7 Simulating the Consequences of Policies That Improve Environmental Conditions 131
Introduction 131
Green Gentrification 132
Different Neighborhoods, Different Outcomes 133
The Model 134
Neighborhood Gentrification 134
The Policy Intervention 135
Simulation Timing 135
Policy Approaches 136
Results 139
Neighborhood Gentrification and Residential Displacement Are Fairly Common in This Simulation 140
Displacement 142
Discussion 149
Conclusion and Policy Implications 150
Appendix: ABM Details 151
Utility Function 151
Land Price and Quality 152
References 154

8 Mitigating Displacement from Green Gentrification: Examining the Role of Housing-Cost Control Measures 157
Soomi Lee and Adam Eckerd
Introduction 157
Policies to Limit Displacement 158
Homeowner Protections 160
Renter Protection 165
Importance of Policy Environment in Assessing Displacement Driven by Green Projects 166
Agent-Based Modeling 167
Base Model Similarities and Differences (Chaps. 5, 7, and 8) 167

Homeowners Versus Renters. 168
Tax Caps 169
Simulation Scenarios 171
Results. 173
Results Based on City Scenarios. 176
Discussion and Conclusion 179
References. 180

9 Distillations and Policy Recommendations 183
Introduction. 183
Insights 184
Insight 1: Green Gentrification and Displacement Are by No Means Certain 184
Insight 2: Some Cities, Places, and Groups Do Need to Worry About Displacement More Than Others, so They Need to Think More Carefully About How to Proceed 186
Insight 3: Those Who Are Displaced Once May Be Displaced Multiple Times 188
Insight 4: Smart Policy Interventions Are Still Possible. 190
Insight 5: What Factors Don't Seem to Matter? 192
Insight 6: Directions for Future Research. 193
Conclusions. 194
Restatement of Some Key Considerations 195
References. 196

Index. 199

About the Authors

Heather E. Campbell is Thornton F. Bradshaw Professor of Public Policy and Director of the Division of Politics and Economics in the School of Social Science, Policy & Evaluation at the Claremont Graduate University in Claremont, California. Her research focuses on urban environmental policy analysis with a specific focus on environmental injustice. She can be reached at heather.campbell@cgu.edu.

Adam Eckerd is an Associate Professor with the School of Public Service at Old Dominion University. He conducts research on organizational and individual decision making in complex settings, particularly as it relates to how risk is assessed and how information is used to manage public and nonprofit programs and policies. Adam's work investigates issues of environmental policy and land use, public participation, and program evaluation. He can be reached at aeckerd@odu.edu.

Yushim Kim is an Associate Professor in the School of Public Affairs and a Senior Sustainability Scholar at Arizona State University. Her research focuses on environmental justice and policy as well as public health service. Her methodological interest is to explore the benefits and shortcomings of developing analytical tools that can structure the complexity of systems, such as agent-based modeling, qualitative comparative analysis, and network analysis. She can be reached at ykim@asu.edu.

Jieun Kim is a doctoral candidate in Public Administration and Policy at Arizona State University. Her research focuses on urban environmental policy and management, with a specific interest in policy design of nature-based solutions for local climate-change adaptation. She can be reached at jkim593@asu.edu.

Soomi Lee is a Professor in the Department of Public Administration and the Director of the Master of Public Administration program at the University of La Verne. Her research interests include state and local public finance and urban/regional economics. She can be reached at slee4@laverne.edu.

Michelle Stuhlmacher is an Assistant Professor in the Department of Geography and GIS at DePaul University in Chicago, Illinois. Her work focuses on urban green space, environmental justice, and community-engaged research. She can be reached at michelle.stuhlmacher@depaul.edu.

Chapter 1
Introduction to Green Gentrification and Our Approach

Abstract With segregation in housing markets a recurring problem in the US, communities populated predominantly by racial and ethnic minorities have tended to be locations for disproportionately more disamenities and disproportionately fewer amenities than other communities; this is the classic Environmental Justice, or EJ, problem. There is considerable evidence on disproportional exposure to environmental disamenities in these communities, with more recent evidence developing with regards to lesser access to environmental amenities such as green spaces. Environmental amenities are not provided in isolation, but are often part of cities' and communities' efforts to renovate and revitalize polluted, segregated, and low-income neighborhoods. Yet the addition of amenities can cause changes to the neighborhood. Without understanding how neighborhoods change in the process of cleaning and greenspace development, it is difficult to tackle environmental injustice issues, for they may be aggravated by the distribution of amenities in urban areas. This book clarifies the separate concepts of gentrification and displacement, considers the city as a complex adaptive system, identifies crucial factors that are involved in neighborhood changes, provides research to gain insight onto the clusters of key factors that may lead to more desired or less desired neighborhood change, and makes policy recommendations.

Introduction

Much research shows that racial and ethnic minorities in the U.S. suffer from environmental injustice, both such that there are *too many dis*amenities in their neighborhoods, and *not enough amenities*. Some of our earlier research indicates that cleaning up brownfields could be a helpful policy intervention, reducing the environmental quality gap while improving or maintaining environmental quality overall (Campbell et al., 2015). However, under some conditions this environmental remediation can also foster "green gentrification" and displacement (e.g., Eckerd et al., 2019), leaving the intended beneficiaries worse off.

H. E. Campbell et al., *Green Gentrification and Environmental Injustice*,
https://doi.org/10.1007/978-3-031-65100-7_1

In this book, we take our cue from the work of Elinor Ostrom, specifically her concept of the configural nature of complex problems: In a complex system it is the *mix* of social conditions that determines the effect of a policy change, rather than the effect of any independent variable in a ceteris paribus world. A city is the very definition of a complex adaptive system: a system of systems with relatively autonomous agents that constantly adapt to changes in the environment. From this perspective, a key problem in addressing urban issues, including both environmental justice (EJ) issues and gentrification, is that we do not know which social conditions are important to the configural mix that will determine the extent to which policy interventions will have positive or negative consequences.

In this chapter we provide an overview of issues surrounding green gentrification and environmental (in)justice, also previewing our analytic approach; the next three chapters provide deeper reviews of these. Then, this book provides novel analytic clarification of different components of the interrelated concepts of economic development, gentrification, and displacement. We use this careful analytic disaggregation, along with complexity-appropriate methodologies (Qualitative Comparative Analysis and Agent-Based Modeling), detailed city-level information, and computational modeling in order to suggest mixes of conditions that constitute key configurations that can explain how cities might achieve cleaning and greening while minimizing justice problems caused by displacing those who have suffered from the existing pollution and therefore, we assert, should benefit from its amelioration.

Gentrification and Displacement

As discussed further in Chaps. 2 and 4, over the time that it has been studied as an urban phenomenon, gentrification has been viewed with different connotations: neighborhood improvement—which can be triggered by economic development—or as loss-inducing compositional shift and displacement. Both people and space are intertwined with the concept of gentrification. While acknowledging the interconnectedness of both of these processes, our approach is to provide a more theoretical explanation, use empirical evidence from case studies and theoretical evidence from computational studies and, subsequently, provide useful insight to policymakers seeking to provide neighborhood reinvestment while reducing displacement. We thus differentiate these concepts based upon the specific processes that are at work. Economic development is desired investment in a community that brings amenities, employment, and other social goods to a community that had previously lacked them. Gentrification occurs as property values increase and as the demographic nature of the community changes in response to these economic development changes (or to other types of changes detailed later), and displacement occurs as gentrification accelerates and rents and property values become too high for a community's original residents to remain in place and they are forced to relocate.

Recent Gentrification and Displacement in the US

According to Hwang and Lin (2016), in the US gentrification is more common since 2000 than it was during the 1990s (or before), particularly in the largest cities. Recent gentrification has been associated with increases in the percent of *college-educated* residents in urban communities, as well as increases in the percent of *White households*, and increases in *rents* (Hwang & Lin, 2016).[1] The National Community Reinvestment Coalition (NCRC) found that seven cities accounted for nearly half of the gentrification nationally from 2000 to 2013: New York City, Los Angeles, Washington, D.C., Philadelphia, Baltimore, San Diego and Chicago (Richardson et al., 2019). The presence of these cities on the list is no surprise, but notice that somewhat more than half of the gentrification nationally during this period was not in these obvious locations. Some suggest that this post-2000 gentrification is influenced by the rise of the Millennial generation as a major labor group because the Millennial generation exhibits patterns of delayed household formation and homeownership. Demographic changes have led to increased representation of nonfamily households, childless families, and young adults. Additionally, with changing patterns of racial prejudices, biases, and inequalities, some of the younger population shows increasing preference for diverse neighborhoods (Hyra, 2016).

While, in its earliest form, gentrification was seen as the replacement of poorer residents with richer ones (Glass, 1964), in the American context, much of the concern with gentrification has focused beyond income-level displacement to racial/ethnic displacement. There are various reasons to be concerned with racial/ethnic displacement, such as loss of cultural elements, including food culture. However, our concern with racial/ethnic displacement is explicitly embedded in our concern with EJ. Since decades of research across a variety of pollutants shows that racial and ethnic minorities disproportionately suffer from degraded and unhealthy environments, our concern with cleaning and greening neighborhoods in order to improve the environment—and reduce the EJ gap—requires that residents who have suffered from pollution be able to stay in an improved neighborhood long enough to gain some of the benefits of cleanup.

All of this noted, it is important to point out that the influx of this new generation or rising middle class who can afford gentrified inner-city housing and amenities may not necessarily directly lead to the displacement of low-income minorities. As we elaborate further below, there are cases in which neighborhoods gentrify—meaning that housing values rise—but minorities are not displaced (Richardson et al., 2019; Gamper-Rabindran & Timmons, 2011).

Some level of change in neighborhoods is always expected, but when changes are driven by factors that communities and governments did not intend, or when communities and governments can mitigate undesirable aspects of neighborhood changes, it is worth examining the issue from a policy perspective.

---

[1] By some conceptions of gentrification, the last is almost definitional.

A Gentrification Framework: Government Intervention

Figure 1.1 presents a framework that can help make sense of government's involvement in gentrification and displacement in neighborhoods as well as larger regions. Government's goal (at least putatively) is to improve neighborhoods via cleanup and/or the addition of amenities including jobs and housing. The center shows that the benefit of neighborhood improvement often comes with the cost of increases in property values and rents. This is an opportunity for those who can afford the change to enjoy the benefits of improvement, but this also forces out those who cannot afford the change from their own preferred neighborhoods.

The crux of the gentrification-related displacement problem is *significantly increased property values and rents in neighborhoods* which may create an affordable-housing crisis. A gentrification-displacement cycle begins, where property values escalate, displacing even more people, until, in the end, the community becomes wealthy[2] but with very little of that wealth being accrued by the original residents—that is, the wealth is not derived within the community, and the wealth that moves from outside the community into the community primarily stays with the in-comers. As we detail more in Chap. 4, however, many of these elements are separable. They do not always go hand-in-hand.

We further elaborate this perspective in Fig. 1.2, in which we distinguish our concepts of space and people in the gentrification context. Useful conceptualization

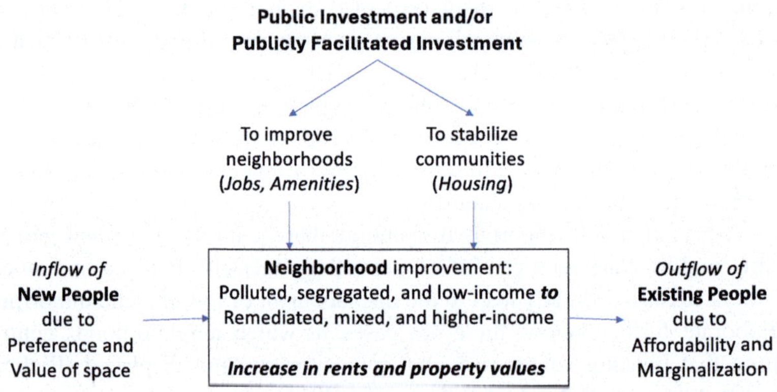

Fig. 1.1 Elements of gentrification and displacement processes triggered by government. (*Source*: The Authors)

[2] Here when we discuss wealth in the community we mean both individual and public wealth.

of neighborhood change is made more difficult by the fact that economic development is clearly related to, but not the same as, gentrification. Economic development can trigger gentrification, but so can other factors. Gentrification itself is often measured by income increase, but it is also strongly tied to physical change, sometimes caused by explicit economic development interventions (including cleanup), and sometimes by individual actions.

Here we conceptualize gentrification as a concept related to the space improvement tied to residents' income status—areas of cities gentrify when the space is upgraded in some way. This may occur because the incomes of those in a neighborhood rise, leading to spatial improvement, or because spatial improvement leads those there to improved employment opportunities, or because spatial improvement leads to the inflow of those with higher incomes.

We view displacement as a concept specifically related to people and not to space, and it goes beyond the random residential movement that occurs constantly. In displacement, as spaces gentrify, people relocate, either because they are priced out due to the inflow of those with higher incomes, or they no longer feel like the neighborhood is a cultural fit. In some cases people have said that they no longer felt welcome in what had been their neighborhood.

In Fig. 1.2, on one axis we use neighborhood household income increase, compared to the average of the metropolitan area, as a proxy for neighborhood improvement, a common approach in gentrification literature (Ellen & Ding, 2016; Hwang & Lin, 2016). Some neighborhoods' incomes may change relative to the metropolitan-area's change during a relevant time period[3] (cells A and C).[4] In the other axis, we present the primary beneficiary of neighborhood changes. When residents who

Neighborhood Change		***Primary Beneficiary***	
		Existing Residents	Newcomers
Gentrification Neighborhood's average income, compared to average of the metro area	Below Average	**A**	**B**
	Significant Increase	**C**	**D**

Fig. 1.2 Conceptual clarification of gentrification and displacement. (Source: The Authors)

[3] It is not clear from our scan of the gentrification literature what the most appropriate temporal context is. It is clear that gentrification and displacement take some amount of time to fully process, but the time horizon in any particular neighborhood or region can be highly variable. We expect that most gentrification and displacement processes will be revealed over a decade. In Chaps. 4 and 7 we further discuss the importance of the time period.

[4] Of course income levels can also decline, but this is not relevant to gentrification; this is more like blight.

predated any investment are the primary beneficiaries of policy (cell C), we may consider the neighborhoods as cases of economic development. Although this is not the terminology usually used in the literature (with occasional exceptions—see Byrne, 2002), we might say that such neighborhoods have been gentrified without (much) displacement. When newcomers in neighborhoods are the primary beneficiaries of improvement, we consider those neighborhoods as having been gentrified with displacement (cell D). In cell D, a neighborhood experiences both gentrification and displacement at the same time.[5]

We see the distinction between cells C and D as crucial for understanding how to reinvest in urban neighborhoods. Virtually every stakeholder involved—and certainly current residents—desires a transition to cell C and not to cell D (Cox et al., 2000). Yet we know that cell-D change often occurs, and the fear of economic development leading to displacement may alter development preferences, perhaps even to the extent that community residents have become wary of efforts to foster investment lest those investments lead to displacement as is the case with cell D (Curran & Hamilton, 2017; Times Editorial Board, 2017). In one case in Chicago a rail-to-trail project failed because of residents' fears of gentrification (Rigolon & Nameth, 2018). Similarly, in East Baltimore a tree-canopy project (which can reduce urban heat island problems) was only partially successful due to residents' suspicions (Rigolon & Nameth, 2018). But, within the context of systemic environmental injustice, this is problematic; such fears may prevent minorities in polluted areas from gaining needed cleanup.

Our interest here is to consider whether, how, and under what circumstances neighborhoods can spur development that enables communities to transition to cell C rather than cell D. To answer the question, we must explore what potential paths tend to lead one way or the other.

Better to guide our thinking as to how to have environmental improvement without gentrification that leads to displacement, in Chap. 4 we further elaborate on what we consider to be such useful analytic distinctions as those presented in Figs. 1.1 and 1.2.

Can There Be Gentrification Without Displacement?

An ongoing question in the gentrification literature is whether gentrification can occur without displacement. Above, and especially in Chap. 4, we detail theoretically how this could occur, but it still begs the question as to whether this decoupling can actually be observed. Richardson et al. (2019), all of the National Community Reinvestment Coalition (NCRC), do provide strong evidence that gentrification can occur both with and without displacement. Figure 1.3, taken from

[5]Though not our concern in this work, we view the change from A to B as typical of the sort of immigration succession that a city such as New York may observe, where an Irish neighborhood becomes Puerto Rican, a German neighborhood becomes Chinese, etc.

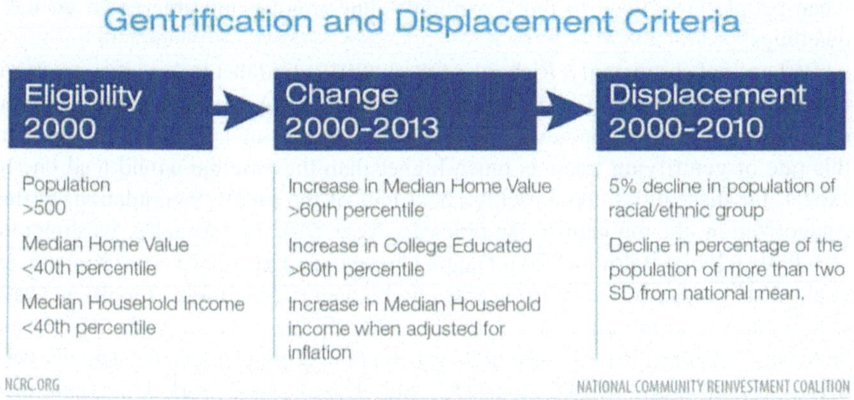

Fig. 1.3 Criteria for gentrification and displacement at census tract level using 2000 and 2010 decennial census and 2009–2013 5-year ACS. (*Source*: Richardson et al., 2019, p. 13; used with permission)

their 2019 report, explains how they conceptualize the different processes. Their word "eligibility" matches the term we will use throughout this book: "gentrifiable." Their study uses nationwide "census data normalized by the longitudinal tract database (LTDB)" (p. 11). Then, they examine two basic attributes: "educational level and economic status of residents, and the value of properties in the neighborhood" (p. 12). Thus, they use indicators of the two different facets identified above: attributes of the people, and attributes of the neighborhood/tract. In their analysis, they only examine tracts with at least 500 residents, and to be "eligible" to be gentrified these must be "below the 40th percentile in both median household income and median house value" (p. 12). Those that then meet the criteria in the "Change" column in Fig. 1.3 (significant percentile increases in home value and college-educated residents, and real increases in income) are considered to have gentrified.

But, in order to define a gentrified tract as also exhibiting displacement, there need to be declines in "minority subpopulation," where "non-Hispanic White, Black, Hispanic and Asian" were analyzed (Richardson et al., 2019, p. 12). Because of our concern with race-and-ethnicity-based environmental justice, their specific concern with racial/ethnic displacement, which they call "cultural displacement," is directly relevant to our analysis and concerns. In this NCRC study, over 11,000 Census tracts met the eligibility criteria, and these represent about 16.7% of urban tracts. Of these, about 1000 (9%) gentrified as defined, and then, of those, about 22% exhibited racial/ethnic displacement (p. 13).

One of the issues in this literature in general is that it is very difficult to assess empirically whether people have been displaced or have simply moved. Americans are quite mobile, and low-income people in particular tend to move at a high rate from all neighborhood types (Desmond, 2016, in Hyra, 2016). In our analyses in our computational-method chapters (Chaps. 5, 7, and 8), we can finesse this by directly observing what causes our artificial agents to move, with displacement defined when people are forced to move explicitly due to not being able to afford their dwellings.

Perhaps not surprisingly, Richardson et al. (2019) find that larger cities are more likely to gentrify, with "CBSAs [Core Based Statistical Areas] with a population of one million or greater containing 79 percent of gentrifying tracts" (p. 14). Further, this rate of gentrifying tracts is much higher than the baseline would lead one to expect, for these cities contain only about half of the total US population. In this categorization, the top gentrifying cities are New York, Los Angeles, Washington, DC, Philadelphia, Baltimore, San Diego, Chicago, and Seattle (p. 16).[6] Richardson et al. (2019) also consider at what rate eligible/gentrifiable tracts actually gentrify within cities, and the outcomes here are rather different. Cities with the greatest "intensity" of gentrification—the highest percent of eligible tracts that actually gentrify—are, in order, Washington, DC, San Diego, New York City, and then Albuquerque, Atlanta, Baltimore, Portland and Pittsburgh (p. 17).

As indicated above, during their study period, and using their criteria, 232 of the 1049 gentrified tracts also exhibited displacement. 187 tracts showed sufficient decrease in Black residents to qualify as displacement, with a total of almost 111,000 Blacks removing from these tracts; 45 tracts showed sufficient decrease in Hispanic residents, with more than 24,000 Hispanics removing from these tracts (Richardson et al., 2019, p. 20). These amounts are not insignificant, but it is important to note that by no means the majority of gentrifiable tracts gentrify, and by no means the majority of gentrified tracts lead to displacement as measured. Thus, it is plausible that we can observe factors that make it more and less likely that green gentrification leads to harmful displacement. Fitzgerald (2010) also finds two cases, one in Denver, Colorado, and one in Portland, Oregon, where transit-oriented development did lead to space improvements but had not led to displacement by the time of writing.

Potential Causes of Gentrification

Variation by city is observed in the average numbers of minority persons displaced in the Richardson et al. (2019) tracts. Hispanics show more variation than Blacks, with between 181 and 1054 Hispanics displaced on average per tract per city, while the range for Blacks is 280 to 935 on average per tract per city (pp. 21–22).[7] The rates of displacement from gentrified tracts also varied significantly by city, from a high of about 78% of gentrified tracts exhibiting Black displacement in Richmond to a low of about 8% in Los Angeles (p. 23). In general, "Cities with the highest levels of black displacement… were concentrated in the South" (p. 22); "Hispanic

[6] At the time of writing, the three largest metro areas were New York City, Los Angeles, and Chicago (Richardson et al., 2019, p. 16).

[7] Not including New Orleans at 1075 because this was a different phenomenon, exacerbated by Hurricane Katrina.

displacement was most intense in the West" (p. 23). African-Americans are generally the largest minority group in the South, and Hispanics in the West. In this book we examine city-level differences in segregatory preferences, which ties in well to these observations.

Another factor that Richardson et al. (2019) believe to be important in displacement is rates of suburbanization, for they see a pattern of greater displacement near central cities: "This study indicates that minority displacement is indeed occurring in many cities, where it is often concentrated in areas near the central business district" (CBD) (p. 29). Compared to Blacks, Hispanics live less often in the downtown core, and Hispanics are less segregated from non-Hispanic whites, as well.[8] Richardson et al.'s (2019) results regarding regional differences in attitudes toward minority groups, and differences in neighborhood similarity, ties in well with concepts we have examined before and examine further in this book (see, e.g., Chap. 5).

In their review of the gentrification literature, Hwang and Lin (2016) identify that the presence of or increases in access to jobs and amenities are associated with increased demand for downtown neighborhoods. However, the evidence on job access is mixed and inconclusive and, as with much urban policy research, suffers from an endogeneity issue that occludes distinguishing whether workers followed jobs or jobs followed workers.

Evidence on the influence of amenities (e.g., theaters, museums, restaurants, grocery stores, personal service establishments, greenspaces, parks) as drivers of gentrification is relatively robust (Hwang & Lin, 2016), but, again, potentially suffers the same endogeneity problem discussed above.

There are other conditional factors that seem to make urban communities more or less attractive. Access to public transit has been shown to make communities more susceptible to gentrification and displacement. Areas that have a distinct architectural quality have been shown to be prone to gentrify, particularly if there is a high level of residential vacancy (Eckerd, 2011). The crime rate has also been shown to be a particularly important consideration—incoming residents (sometimes called "incomers," "urban pioneers"—or urban "colonizers" by Mitchell, 2019) appear to be willing to live with some amount of urban crime, but there seems to be a point at which there is too much crime for gentrification to begin. All in all, we tend to see that when the factors that we might associate with a high quality of life (i.e., safety, amenities, convenience, aesthetics) are aligned in an urban area, it tends to be more prone to gentrify than when those factors are missing or more limited.

The overall policy challenge is that these quality-of-life factors are exactly the sorts of things that urban governments want to provide to communities—and we certainly do not want the advice to government to be to avoid improving quality of life.

---

[8] Richardson et al. (2019) report that the dissimilarity index during their study period is 0.6 while it is 0.4 for Hispanics (p. 24).

Mitigators of Gentrifier Inflow

As indicated in Fig. 1.1, community and government investment in neighborhoods can contribute to favorable environments for influx, but there are also factors that limit the demand that arises with community changes. Most gentrification research in the US is focused on a handful of cities that are well known for being hotbeds of gentrification (e.g., Gould & Lewis, 2017). In these cases high socio-economic status (SES) households that already live in the region move into these rapidly redeveloping central cities; this creates new demand for housing supply, but often housing supply is not flexible, which leads to price increases. It may be that the increasing international fiscalization of housing, in which it is used as a store of value rather than as a dwelling, exacerbates these price increases. In cases where the supply of housing is ample, there may be less displacement. In cases where housing supply is tightly controlled or restricted by local rules and policies, there may be more displacement because of rising housing costs and rents. The cities that have been studied most extensively with respect to gentrification and displacement tend to have land constraints that limit housing availability (like New York, Boston, or San Francisco) and/or strict development policies that limit housing supply, like Portland's urban growth boundary.

Contrariwise, gentrification has also been slower to take hold in cities that do not fit this profile, particularly places where transportation costs and commute times are low, like many mid-sized cities, or where there are few land or development constraints (like Houston or Phoenix). This work suggests that there is a trade-off between housing costs and transportation costs—where these costs are low, higher wealth residents may well prefer suburban or exurban living in general, limiting the displacement potential of inner-city investments. In contrast, when commute times are long or costly, higher wealth residents may prefer living closer to work. This fits well with certain theories of urban development (e.g., Bluestone et al., 2008).

Mitigators of Displacee Outflow

Government's role as outflow mitigator has also been noted by several scholars. Governments at different levels utilize different policy levers. Focusing on income displacement, federal policies often aim at increasing the supply of affordable rental housing (Hyra, 2016; O'Regan, 2016) or preserving affordable housing. More frequently, federal policy encourages localities to employ a range of tools, proactively and broadly. Lubell (2016) discussed six different categories of local policy levers to respond to housing affordability crises: (1) preserve existing affordable rental units, (2) help long-time residents who wish to stay in the neighborhood, (3) ensure that a share of new development is affordable, (4) harness growth to expand financial resources for affordable housing, (5) create incentives for developers of affordable housing, and (6) facilitate the acquisition of sites for affordable housing. Note that five of these "different categories of local policy levers" involve affordable housing. Affordable housing is often, in fact, promised with redevelopment, but promises are not always met, and in the US context even the promised amounts are usually significantly less than the demand (see Gould & Lewis, 2017).

Another way to mitigate outflow may be to create policies to forestall culturally based displacement such as holding cross-race, cross-culture connection events in communities, and instituting low-income civic engagement programs (Hyra, 2016)—or, stated more simply, fostering social capital between long-time residents and the people moving in. Some considerations in regard to the importance of social capital accord with findings reported in Taylor (2014, p. 83) regarding the ability to resist the placement of disamenities in the first place. As cited in Taylor (2014), "Gayer (2000) reports a similar finding—he found that polluters placed their facilities in communities less likely to engage in collective action" (p. 83).

Part of the challenge, however, is that these issues—all of the factors that are associated with gentrification and displacement—potentially exhibit interdependence and thus endogeneity, making it difficult to discern what factors need to be considered when crafting policy that encourages economic development without displacement. This, along with the fact that cities are Complex Adaptive Systems (as we explain in Chap. 3), suggests that the process of gentrification and displacement is an example of a wicked problem,[9] which requires a different orientation for addressing the problem—notably a recognition of the interconnectedness of the factors that influence what happens. Rather than conceiving of these factors individually, we need to consider the mix, or the configurations, of different factors and how the process of gentrification plays out in different settings.

Policy Actions

Thus, we need to consider how quality of life improvements can be made, while mitigating the potential for displacement. Neighborhood improvement, especially via pollution remediation and addition of amenities, safety, and accessibility, are facilitated by government, particularly via state and local investment. Owing to the prior several decades of disinvestment, many urban policies of the 1990s were explicitly aimed at bringing the middle class back to the urban core (Hyra, 2016)—for a variety of reasons, from increasing tax revenues, to goals of increasing income "social mix" due to beliefs that concentrated poverty is bad for the poor. Some revitalization projects funded by governments demolished public housing in

[9] Please note that the term "wicked problem" has to do with the nature of the problem—that it is difficult to solve due to its complexity—and in no way implies that populations within gentrifying neighborhoods are wicked. We use this term as it was introduced in the '60 s and has been used in planning and policy since (see, e.g., Churchman, 1967, Rittel & Webber, 1973).

neighborhoods and provided opportunities for the development of new market-rate housing (Wyly & Hamel, 1999; Goetz, 2011).[10]

For most of the last 30–40 years, governments have preferred to utilize free-market solutions to urban problems, promoting business-friendly development and public-private partnerships (Hackworth, 2007) over more direct regulatory approaches. These efforts are more likely to have greater marketized returns (including greater escalations of both rents and property taxes), but these approaches also have often promoted gentrification in or near minority neighborhoods that middle-class residents once avoided. Historical preservation, business improvement districts, zoning and land-use changes, tax-increment financing practices, ordinances imposed on public space, and the beautification of public spaces have shifted the amenity value of neighborhoods (Michell, 2003; Ward, 2007; Weber, 2002; Wilson & Mueller, 2004; Zukin et al., 2009). To attract the "creative class" (Florida, 2002) to downtown neighborhoods, things like parks and bike shares were provided. Redeveloped housing and hip amenities like espresso shops and upper-class amenities like high-end restaurants facilitated the influx of those who are willing to and can afford urban living in the areas with increased housing value as well as rents—often at the cost of displacing the poor who had settled in the urban areas disinvested since the 1960s.

In the US in the past 20 years, concerns about displacement have become more of a primary driver of urban politics, with a resultant concern about how public policy has fostered gentrification. Policies now often do focus on recommended issues like maintaining or requiring affordable housing, and cultural preservation, but there is little indication that these efforts have done much to stem the tide of gentrification and displacement—if anything, the twenty-first Century has been characterized by a continuing reliance on market-oriented approaches and more gentrification in more types of cities (Kadi & Ronald, 2014).

However, gentrification does not always occur, and it does not always lead to displacement; these sets of circumstances may offer some clues as to how policy may encourage development without displacement. Gamper-Rabindran and Timmins (2011) perform a Census-block-level study of a national sample of Superfund sites that were cleaned up, comparing them to others that were not cleaned up. They find clear evidence of in-migration by richer and better-educated residents, but not clear displacement of minorities. Interestingly, they observe an increase in density; increasing residential density is one way to enable in-migration without displacement. Considering density, it is interesting to note that, recently, California has created a state-wide policy allowing for increased residential density. In our analyses in Chap. 5 we include density differences.

[10]Although it is worth noting that the development of these public housing projects fostered the blight that is often a precondition for redevelopment and gentrification to occur. See S. Mitchell (18 March 2019) for an example in Richmond, Virginia, where 60% of public housing is in one of 9 districts. Prince (18 March 2019) also argues this happened in DC, with intentional neglect followed by "improvements" that led to displacement.

Street-Level Insights

As noted above, in an NCRC report, Richardson et al. (2019) find that, analyzing the whole U.S. over 13 years, about 9% of gentrifiable tracts gentrified, and of those 22% also exhibited displacement as they rather strictly measured it. What can we learn from these cases? As addenda to their statistical analysis, Richardson et al. provide case discussions, provided by local experts, for the cities of Philadelphia, Pennsylvania; Portland, Oregon; Richmond, Virginia; and Washington, DC (https://ncrc.org/gentrification/).

According to local applied-equity-policy expert Goughnour, Portland, a city that started with less than the national average of African American residents, has seen gentrification displacement lead to even fewer African Americans, and those remaining also live more scattered (Goughnour, 2019). Emphasizing the gentrification trends we have seen since 2000 in US cities, in Portland an urban renewal plan was instituted in 2000 and included "investment in the heart of the African-American community" (Goughnour, 2019). Portland is now often considered one of the most livable cities in the US, but "Since 2000, all ten of the city's majority minority neighborhoods have become majority white" (Goughnour, 2019). Stressing ongoing concern in the literature with costs of displacement, displaced residents moved to "where the sidewalks, grocery stores and parks grow sparse, and access to public transit is limited" (Jones, as quoted in Goughnour, 2019). Considering Fig. 1.1, above, governmental intervention appears to have succeeded regarding the left side of the figure—spaces were improved—but to have failed regarding the right side, where community stability is a goal based in the idea that those who have suffered from underinvestment, pollution, and blight should also benefit from investment that leads to improvement.

According to non-profit manager Mitchell (2019), as a former capital of the Confederacy, Richmond's history is steeped in racism and race-based social injustice. Mitchell's discussion presents a personal cri de cœur as she discusses her own repeated displacement and her sense of culture loss. Mitchell and her husband were involuntarily displaced 4 times in 5 years, and "Each time we moved, we paid more in rent for fewer square footage" (Mitchell, 2019). In Chaps. 5 and 7 we analyze how *often* those displaced may be displaced.

According to activist and cultural anthropologist Prince (2019), gentrification in Washington, DC, has included the privatization of public property, an extreme version of neo-liberal marketization that seems particularly startling in "the country's city." Prince also explicitly discusses the difference in what high-income residents may prefer compared to low-income residents, contrasting "dog parks, bike lanes, condominiums and expensive restaurants" with "truly affordable housing, safe and reliable child care, food justice, the ability to attend houses of worship and the unencumbered use of public spaces" (Prince, 2019). Prince also discusses costs of displacement, including additional errands and "the loss of neighbors and social networks" (Prince, 2019). She presents goals, not policies, but her view that lower-income residents and higher-income incomers may have different amenity preferences fits well with the "just green enough" approach to green gentrification, and we discuss this approach in Chap. 2 and analyze such issues in Chap. 7.

For Philadelphia, several policies are discussed by nonprofit policy director McConnell, with the featured anti-displacement policy approach being policies to help homeowners who face high property taxes (McConnell, 2019). Low-income senior homeowners and long-time owners may have property taxes frozen, though there are not similar programs for renters. But, as of the time of writing, gentrification plus displacement had not been solved in Philadelphia. One problem is that their tax-relief programs are confounded by another policy that provides "10-year Property Tax Abatement offered for all new construction" (McConnell, 2019). Up until now, the gentrification literature, in contrast with the public finance literature, has not featured consideration of property-tax abatement measures. Could such a policy, implemented cleanly, reduce the likelihood of gentrification displacement in some cases? We examine this issue in Chap. 8.

These cases resonate with the gentrification and displacement literature, and they offer some hints of factors to consider, but they all seem disappointing, not offering much promise that any of these four cities have found a particularly effective policy intervention in their settings. Thus, the question remains: What interventions seem likely to work within which settings? We hope the analyses in this book will shed some light.

Discussions and Conclusion

The "markers of gentrification include neighborhood changes in physical, demographic and cultural characteristics" (Hwang & Lin, 2016, p. 10). In the United States, displacement has been viewed as an issue of class, an issue of race, or as both. For the most part, the literature on gentrification is concerned primarily with the displacement of "low-income minorities," including, sometimes, LGBTQ+ who used to live in inner-city downtown neighborhoods (e.g., Paul Solman's interview with Richard Florida in the PBS NewsHour June Solmon (2017)). There is some limited research that considers white displacement, and also the idea of rural gentrification (Ghose, 2004). Neighborhood improvements such as increased job access and added amenities (e.g., cleaner land, more commercial services, greenspace) might become available thanks to state or local investments to improve urban neighborhoods, and especially for the benefit of low-income minorities. However, it has often been the case that these benefits are actually enjoyed by new residents that are moving into these communities—many of whom would not have moved into the communities before the investment occurred. In these cases, the costs of pollution, neglect, and blight were borne by existing residents while benefits are enjoyed by newcomers.

Residential displacement may also have political and cultural consequences for social groups who used to live in the inner-city downtown neighborhoods.

Residential displacement can be related to a loss of minority political representation in a city as newcomers engage with and eventually supplant leadership of political institutions (Hyra, 2016). It can also be related to a loss of cultural aspects as newcomers advocate for amenities and services that fit their definition of community improvement (Hyra, 2016) but which may not reflect the preference of existing residents. The consequences of political and cultural displacement can include intense social tensions, limited meaningful social interactions between longtime residents and newcomers, and microlevel segregation (Hyra, 2016). Loss of community and loss of culture in economically transitioning neighborhoods are non-trivial issues that can be particularly problematic for low-income residents who rely on social networks to fill in for missing economic resources (Hyra, 2016).

The concerns around gentrification and displacement touch on many different aspects of peoples' lives. First, there is the simple recognition that benefits may be flowing to people who already receive substantial political and policy advantages rather that to those who need more support. Second, in addition to the somewhat less-monetarily-tangible considerations noted above due to the loss of community, there are tangible monetary losses in terms of increasing rent or property-tax burdens and relocation costs. Third, there are concerns around the trend of where displaced residents are moving—such as to more-polluted neighborhoods or to older inner-ring suburbs that lack the tax bases of central cities and may be even less able to provide the services that poor people need (Fullilove & Wallace, 2011). Lastly, it is also an environmental injustice problem that adds burdens to already distressed social groups suffering from disproportionate environmental disamenities.

The gentrification literature suggests that one way that we can manage or address environmental injustice problems in urban areas is to understand how cities can achieve just growth so that already depressed social groups are not forced to leave their improved community (Benner & Pastor, 2012). On the other hand, the challenge of gentrification and displacement is that neighborhood changes in urban areas cannot be understood or influenced well by focusing on any single factor. As noted by other scholars, there are many factors to consider:

> Demand for certain neighborhoods, especially in central cities, has increased since 2000. Although we have some evidence that the presence of jobs and amenities is spurring gentrification in some neighborhoods, we have less understanding of the causal role of redevelopment decisions, supply constraints, or shifting preferences (Hwang & Lin, 2016, p. 21).

Wicked problems cannot be easily fixed with narrow policy solutions because they often lead to further unintended consequences because the interacting nature of different social and environmental factors does not follow linear processes. Taking this into account leaves us with a problem in addressing gentrification displacement: We do not know *which* social conditions are important to the configural mix that will determine the effects of policy interventions. As reported above, Richardson et al.'s (2019) work suggests at least three factors: nearness to the CBD, region of the country (which we take to indicate local rates of prejudice or segregatory preference), and residential segregation.[11]

But, there are multiple interdependent things going on that make some places more or less prone to gentrification and to displacement. Policy also needs to be understood to take place within the confines of that interdependent system that accounts for many of the different mechanisms that were described above and discussed by us—and others that will be reviewed in succeeding chapters. Since this is the task at hand, in this book we propose to give up somewhat the granularity of variables as a way of obtaining or improving our knowledge. Instead, we take the stance that knowledge on complex problems can be improved by identifying the interdependent structures behind the complex problems and rethinking the roles and effects of policy as a more nuanced influence to the system, rather than as a mechanistic policy lever. In service to these goals, the next chapters overview important literatures, further explicate the importance and implications of viewing cities as complex adaptive systems, explore the conceptual separability of gentrification and displacement, and then use complexity-relevant methods to analyze which policies might influence outcomes within different urban configurations.

References

Benner, C., & Pastor, M. (2012). *Just growth: Inclusion and prosperity in America's metropolitan regions*. Routledge.

Bluestone, B., Stevenson, M. H., & Williams, R. (2008). *The urban experience: Economics, society, and public policy*. Oxford University Press.

Byrne, J. P. (2002). Two cheers for gentrification. *Howard Law Journal, 46*(3), 405-432.

Campbell, H. E., Kim, Y., & Eckerd, A. (2015). *Rethinking environmental justice in sustainable cities: Insights from agent-based modeling*. Routledge.

Churchman, C. W. (1967). Wicked problems. *Management Science, 14*(4/December), B-141–B-142.

Cox, A. M., Alwang, J., & Johnson, T. G. (2000). Local preferences for economic development outcomes: Analytical hierarchy procedure. *Growth and Change, 31*(3), 341–366. https://doi.org/10.1111/0017-4815.00132

Curran, W., & Hamilton, T. (Eds.). (2017). *Just green enough: Urban development and environmental gentrification*. Routledge.

Desmond, M. (2016). *Evicted: Poverty and profit in the American city.* Crown. As cited in Hyra (2016).

Eckerd, A. (2011). Cleaning up without clearing out? A spatial assessment of environmental gentrification. *Urban Affairs Review, 47*(1), 31–59. https://doi.org/10.1177/1078087410379720

Eckerd, A., Kim, Y., & Campbell, H. (2019). Gentrification and displacement: Modeling a complex urban process. *Housing Policy Debate, 29*(2), 273–295. https://doi.org/10.1080/10511482.2018.1512512

Ellen, I. G., & Ding, L. (2016). Guest editors' introduction: Advancing our understanding of gentrification. *City, 18*(3), 3–8. http://www.jstor.org/stable/26328270

[11] Note that segregatory preferences influence segregation levels, but the two are not the same because in a complex adaptive system people end up living with more segregation than they prefer.

Fitzgerald, J. (2010). *Emerald cities: Urban sustainability and economic development*. Oxford University Press.

Florida, R. (2002). *The rise of the creative class: And how it's transforming work, leisure, community, and everyday life*. Basic Books.

Fullilove, M. T., & Wallace, R. (2011). Serial forced displacement in American cities, 1916–2010. *Journal of Urban Health, 88*(3), 381–389. https://doi.org/10.1007/s11524-011-9585-2

Gamper-Rabindran, S., & Timmins, C. (2011). Hazardous waste cleanup, neighborhood gentrification, and environmental justice: Evidence from restricted access census block data. *American Economic Review, 101*(3), 620–624. https://doi.org/10.1257/aer.101.3.620

Gayer, T. (2000). Neighborhood demographics and the distribution of hazardous waste risks: An instrumental variables estimation. *Journal of Regulatory Economics, 17*(2), 131–155. As cited in Taylor (2014).

Ghose, R. (2004). Big sky or big sprawl? Rural gentrification and the changing cultural landscape of Missoula, Montana. *Urban Geography, 25*(6), 528–549. https://doi.org/10.2747/0272-3638.25.6.528

Glass, R. (1964). *London: Aspects of change*. MacGibbon and Kee.

Goetz, E. (2011). Gentrification in black and white: The racial impact of public housing demolition in American cities. *Urban Studies, 48*(8), 1581–1604. https://doi.org/10.1177/0042098010375323

Goughnour, C. (2019, March 18). *Portland, Oregon: Displacement by design*. NCRC. https://ncrc.org/gentrification-portlandor/

Gould, K. A., & Lewis, T. L. (2017). *Green gentrification: Urban sustainability and the struggle for environmental justice*. Routledge.

Hackworth, J. (2007). *The neoliberal city: Governance, ideology, and development in American urbanism*. Cornell University Press.

Hwang, J., & Lin, J. (2016). What have we learned about the causes of recent gentrification? *City, 18*(3), 9–26. https://www.jstor.org/stable/26328271

Hyra, D. (2016). Commentary: Causes and consequences of gentrification, and the future of equitable development policy. *City, 18*(3), 169–178. https://www.jstor.org/stable/26328279

Kadi, J., & Ronald, R. (2014). Market-based housing reforms and the 'right to the city': The variegated experiences of New York, Amsterdam and Tokyo. *International Journal of Housing Policy, 14*(3), 268–292.

Lubell, J. (2016). Preserving and expanding affordability in neighborhoods experiencing rising rents and property values. *City, 18*(1), 131–150. https://www.jstor.org/stable/26328276

McConnell, B. (2019, March 18). *Learning from our mistakes: Anti-displacement strategies in Philadelphia*. NCRC. https://ncrc.org/gentrification-philadelphia/

Mitchell, K. (2003). Monuments, memorials, and the politics of memory. *Urban Geography, 24*(5), 442–459. https://doi.org/10.2747/0272-3638.24.5.442

Mitchell, S. (2019, March 18). *In Richmond, Virginia, gentrification is colonization*. NCRC. https://ncrc.org/gentrification-richmondva/

O'Regan, K. M. (2016). Commentary: A federal perspective on gentrification. *City, 18*(3), 151–162. https://www.jstor.org/stable/26328277

PBS News Hour. (2017, June 1). *Has urban renewal caused a crisis of success?* Interview of Richard Florida by Paul Solman. https://www.pbs.org/newshour/show/urban-revival-caused-crisis-success

Prince, S. (2019, March 18). *Washington D.G.: The district of gentrification*. NCRC. https://ncrc.org/gentrification-dc/

Richardson, J., Mitchell, B., & Franco, J (2019, March 19). *Shifting neighborhoods: Gentrification and cultural displacement in American cities*. NCRC. https://ncrc.org/gentrification/

Rigolon, A., & Németh, J. (2018). A QUality INdex of Parks for Youth (QUINPY): Evaluating urban parks through geographic information systems. *Environment and Planning B: Urban Analytics and City Science, 45*(2), 275–294. https://doi.org/10.1177/0265813516672212

Rittel, H. W. J., & Webber, M. M. (1973). Dilemmas in a general theory of planning. *Policy Sciences, 4*(2/June), 155–169.

Solman, P. (Host). (2017, June 1). Has urban renewal caused a crisis of success? [Video Interview with Richard Florida]. In *PBS News Hour*. PBS. https://www.pbs.org/newshour/show/urban-revival-caused-crisis-success

Taylor, D. E. (2014). *Toxic communities: Environmental racism, industrial pollution, and residential mobility*. New York University Press.

Times Editorial Board. (2017, July 20). *Editorial: Boyle Heights anti-gentrification activists hurt their cause by making it about race, rather than economics*. https://www.latimes.com/opinion/editorials/la-ed-gentrification-boyle-heights-race-20170721-story.html

Ward, K. (2007). "Creating a personality for downtown": Business improvement districts in Milwaukee. *Urban Geography, 28*(8), 781–808. https://doi.org/10.2747/0272-3638.28.8.781

Weber, R. (2002). Extracting value from the city: Neoliberalism and urban redevelopment. *Antipode, 34*(3), 519–540. https://doi.org/10.1111/1467-8330.00253

Wilson, D., & Mueller, T. (2004). Representing "neighborhood": Growth coalitions, newspaper reporting, and gentrification in St. Louis. *The Professional Geographer, 56*(2), 282–294. https://doi.org/10.1111/j.0033-0124.2004.05602011.x

Wyly, E., & Hammel, D. (1999). Islands of decay in seas of renewal: Housing policy and the resurgence of gentrification. *Housing Policy Debate, 10*(4), 711–771. https://doi.org/10.1080/10511482.1999.9521348

Zukin, S., Trujillo, V., Frase, P., Jackson, D., Recuber, T., & Walker, A. (2009). New retail capital and neighborhood change: Boutiques and gentrification in New York City. *City & Community, 8*(1), 47–64. https://doi.org/10.1111/j.1540-6040.2009.01269.x

Chapter 2
The Environmental Injustice Problem and Its Links to Gentrification and Displacement: Literature Review

Abstract Racial and ethnic minorities in the US have suffered both from having little access to amenities, and from living close to disamenities. Much evidence on disproportional exposure to environmental disamenities has been accumulated, and evidence is emerging that there is also disproportionality regarding access to environmental amenities such as parks and greenspace. These are environmental justice problems. However, attempts to reduce environmental injustice are not provided in isolation, but are embedded in cities' and communities' efforts to renovate and revitalize polluted, segregated, and low-income neighborhoods—and are also embedded in local power dynamics and urban growth-machine politics. Trying to tackle environmental injustice without understanding neighborhood changes that result from pollution remediation and/or the addition of amenities can lead to unintended consequences that aggravate social injustice more broadly. Here we briefly overview issues of environmental injustice, gentrification and displacement, and how these issues are related. As part of this we review the just-green-enough literature.

Introduction

For more than 10 years, separately and together, the authors of this book have been doing research on Environmental Justice (EJ)—or Environmental Injustice (EI) as it really should be known since we have decades of evidence that racial and ethnic minorities in the US suffer disproportionately from pollution, even controlling for factors such as poverty and wealth. In our book on *Rethinking Environmental Justice*, our analysis indicates that a beneficial policy to reduce environmental injustice would be to clean up brownfield sites (Campbell et al., 2015). This approach was estimated to reduce the environmental quality gap between minorities and majorities by improving environmental quality for minority groups without reducing it for majority groups.

In deciding what brownfields to clean up, there are competing criteria. Some use a social efficiency (neoliberal) frame to argue that it makes sense to start with sites that have greater (market) value. Others use an equity/justice frame to argue that we

H. E. Campbell et al., *Green Gentrification and Environmental Injustice*,
https://doi.org/10.1007/978-3-031-65100-7_2

should start by focusing on brownfields in minority communities because they have suffered disproportionately from environmental insult. Others use a public-health frame to argue that we should start with those sites that represent the greatest threat to human health. We personally prefer the latter two approaches as both more accurate depictions of the policy problem and likely more effective bases for solutions, but in any specific city political-economic realities might tend toward the first approach. So, we were pleased when our analysis indicated that it didn't matter much which decision criterion dominated because all led to improvement as measured by a reduction in the environmental-justice gap. Thus, our policy recommendation was clear: clean up brownfield sites, starting with the ones you can start with in your city's political-economic circumstances.

Yet, after this conclusion we began to be concerned about what comes next: the possibility of "green gentrification" and the possibility that cleaning up polluted sites might actually lead to worse outcomes for those already present in the neighborhood (Pearsall & Anguelovski, 2016). "Green gentrification" is the term used when cleaning and/or greening in a neighborhood causes rising prices of land and housing (Dooling, 2018) that eventually lead to displacement of existing residents. As do others, we define environmental cleaning as a type of greening even if greenspace is not specifically added. In other words, simply cleaning up a brownfield site is "greening" in this definition, even if nothing specifically green is placed upon the site.

Given the general mobility of Americans (and even greater mobility of the poor),[1] if we assume that housing markets are reasonably efficient, then people must be at best indifferent to the other neighborhoods where they don't live and, in many cases, prefer their current location to others—especially because most moves are local (Frost, 2020). Being forced out of their current location into another means that they must be worse off. Even if they are indifferent between the new location and the original one, search costs, moving costs, and loss of social capital must reduce their utility.[2] As noted by Richardson and colleagues, displacement itself often has real and immediate consequences:

> Often, they must accept longer commutes and a disruption of the support structures provided by their old neighbors and family. In these cases, gentrification is understood as the terminal stage of exclusion of minority (usually black) residents from affordable housing inside the city (Richardson et al., 2019, p. 7).

In some cases, those displaced are not indifferent, and may be forced into even more polluted areas than the ones they leave. This is the very opposite of our intent and

---

[1] It is worth noting that American mobility has reduced in recent years, with a decades-long declining trend. In the 1980s, almost 20% of Americans had moved in the last year, but by 2020 only 10% had moved in the last year. But, among renters 25% had moved in the last year (Frost, 2020). Americans still move at the highest rate in the OECD (tied with Australians), with 40% moving within 5 years (OECD, 2021).

[2] Though, in the context of a deteriorated city near Tel Aviv in Israel, Levine and Aharon-Gutman (2023) find that social capital in such cities, where resident are socioeconomically disadvantaged and buildings are run-down, may already be nearly non-existent.

goal with respect to our environmental justice research and our policy recommendation to clean up brownfields, which we have pursued because we want to identify policies that can reduce environmental injustice, particularly in US cities.

Some scholars note that brownfield areas may be especially susceptible to green gentrification. For one thing, "environmental cleanup may allow for the realization of the rent gap (Smith, 1996), the difference between actual and potential ground rent that makes gentrification so profitable" (Curran & Hamilton, 2017, p. 15). And, empirical evidence indicates that in some cases price increases can be significant. Reviewing work by Capps (2014) and Maldonado (1996), Tait states that "housing prices may increase anywhere from 4.9 to 32.2 percent" based on brownfield cleanup alone (2015, p. 11).

It is worth noting that the situation is at least somewhat different for renters and owners. In states such as California, due to the famous Proposition 13 (which limits increases in property taxes on continuing owners), owners are less likely to be forced out due to rising property taxes; Lee (2023) finds that there are 19 states and DC which have such assessment growth limits and that evidence indicates that these reduce neighborhood change. In other US states this is not necessarily the case, and extreme increases in property taxes may drive out owners. Yet, even when owners are forced out they are at least somewhat compensated via the increased value of the property they sell—though Levine and Aharon-Gutman (2023, p. 777) argue that even voluntary sales involve displacement because owners can no longer "improve their living conditions within the neighborhood," which especially makes sense if one concern is social capital. But "L. Freeman (2006) asserts that homeowners stand to benefit from gentrification. He demonstrates that some Black homeowners do benefit from gentrification" (Taylor, 2014, p. 277), and Levine and Aharon-Gutman (2023) find that some lower-middle-class owners actually seek gentrification as a means of wealth creation.

On the other hand, displaced renters seem only to lose. They bear search and moving costs and do not carry any new wealth with them as they go. Further, minority renters are likely to be disproportionately poor compared to owners, so such displacement is likely to most harm those who have the fewest social resources. Gould and Lewis, early writers on green gentrification (2017), discuss evidence that, when cleanness/greenness of a neighborhood increases, the average wealth of neighborhood residents increases, and also that an increase in greenness is associated with a decline in people of color (p. 31). They also point out that of course power, wealth, and residential segregation are all interlinked (p. 31). Capps, reviewing a National Bureau of Economic Research (NBER) study, also says that "neighborhoods near brownfield sites that undergo cleanup tend to grow whiter" (2014, first page).[3]

We see cleanup as part of the solution to legacies of pollution for lower-income and minority urban residents, and gentrification displacement as a problem that can prevent this solution from actually helping. As Curran and Hamilton say (2017,

---

[3] It is worth noting that a neighborhood can grow Whiter without displacement if density increases.

p. 41), "From a reparations and representational justice perspective, those communities, both workers and residents, who suffered through decades of toxicity, are precisely the communities who should be served by new green spaces." But, if existing residents are forced out, and those who move in are richer and Whiter and therefore less likely to have been in polluted areas before, this cannot happen.

Nonetheless, concern with green gentrification should not blind us to the fact that *not* cleaning/greening a neighborhood does not guarantee that the utility of that locale stays the same for residents. Gould and Lewis (2017) point out that environmental bads attract more environmental bads, so once a community contains environmentally undesirable uses then, by standard market-influenced political-economic behavior, it is likely to attract more undesirable uses. Similarly, one reason that green gentrification occurs is because adding something good to a neighborhood attracts more good things. As stated by Gould and Lewis, "Due to the intersection of processes reproducing environmental injustices and housing segregation, and the operation of markets and elite actors, the creation or restoration of an environmental good—greening—will tend to increase racial and class inequality and housing segregation, and will decrease environmental justice (National Environmental Justice Advisory Council, 2006)" (2016, p. 36); this is exactly what we fear and examine in this book.

Through the research presented herein, we develop situation-relevant policy suggestions that can reduce the likelihood of green gentrification displacement. As stated by the National Environmental Justice Advisory Council: "'no population should consistently pay a disproportionate price for the cleanup and revitalization of neighborhoods in which they live' (National Environmental Justice Advisory Council, 2006, p. 19)" (Gould & Lewis, 2017, p. 14). Those present in a polluted neighborhood have borne costs—including health costs—of the pollution, and they should receive benefits of the cleanup. Those benefits especially should not accrue only to those who are already better off. As Bell (2004) argues, "a Rawlsian conception of 'justice as fairness' can be readily extended to include environmental justice concerns" (Walker, 2012, p. 44).

Many scholars of green gentrification approach it from an income-class perspective rather than an EJ perspective, and so they are most concerned with class displacement. For example, Derickson et al. (2021) state an urgent question at which those concerned with green gentrification are worrying: "how can we ensure that residents in *low-income* neighborhoods who have suffered from decades of disinvestment are able to share in the benefits of future green investment?" (2021, p. 1, emphasis added), and the only minority distinction that Levine and Aharon-Gutman (2023) make regards international immigrants. The EJ perspective keeps us focused on pollution effects more than disinvestment, and we must not lose sight of the importance of race and ethnicity. To restate Derickson et al.'s (2021) question in the EJ frame: How can we ensure that residents in minoritized neighborhoods who have suffered from decades of disinvestment and pollution-caused health harm are able to share in the benefits of future green investment?

Our focus is improving EJ, health, and avoiding green gentrification displacement as a means of preventing those who have suffered the costs from being moved away from the benefits of cleaning and greening. As Walker points out (2012, p. 176), “A park in this context can be seen as a form of community-scale compensation for having lived with a concentration of pollution—a balancing of a bad with a good.” In order for this to happen, the potential beneficiaries of the greening must be able to stay and reap the good. Further, as pointed out by Levine and Aharon-Gutman (2023), the value of staying in a gentrifying neighborhood may include even more benefits than those caused directly by cleaning and greening if “social mix” (the physical admixture of income-class groups) provides benefits suggested by some of the social-mix literature (for an overview of the social mix literature, see Koo, 2021). In this context, it is also interesting to note that one example of “gentrification with justice,” in urban Atlanta, includes an explicit policy never to place affordable housing units side-by-side (Tait, 2015, p. 19).

In order for any of these benefits to be likely, however, there must be at least some public actors of good will who want policy that allows for green improvements without displacement. Fortunately, both Gould and Lewis (2017) and Derickson et al. (2021) find evidence that such actors do exist within the US urban policy process. Further, our own years of teaching in Public Administration departments leads us to believe that there are public administrators who strive for public good.

A Brief Overview of EJ Literature

The history of Environmental Justice advocacy and analysis in the US is now around 40 years old, and has been reviewed numerous times. Nonetheless, we briefly review it again here both to orient those who are not already familiar with this literature, and also to contextualize our concerns.

It is reasonable to state that Professor Robert Doyle Bullard and his attorney spouse, Linda McKeever Bullard, are the father and mother of contemporary Environmental Justice scholarship. As is often told, in 1979 Linda Bullard was asked to represent Margaret Bean and others, who argued that a permit to place a solid waste facility in a census tract that was approximately 60% minority residents was racist. As part of the court proceedings, Linda Bullard, Esq., asked Dr. Robert Bullard, PhD, to perform statistical analysis of the placement of solid waste sites in Houston. The judge ruled that the plaintiffs did not establish “purposeful racial discrimination” and so their preliminary injunction was denied. However, afterward Bullard wrote and published his article “Solid Waste Sites and the Black Houston Community” (1983), and a new field of academic research was born (Taylor, 2014, pp. 106–109). This court case and the research that Bullard performed as an expert witness triggered at least 40 years of study on what came to be known as Environmental Justice—with environmental *in*justice mostly found.

Another famous early analysis was *Toxic Wastes and Race in the United States*, a study performed by the Commission for Racial Justice of the United Church of Christ (Commission for Racial Justice, 1987). This early study considered “African

Americans, Hispanic Americans, Asian Americans, Pacific Islanders and Native Americans" (Commission for Racial Justice, 1987, p. xi). As has been found numerous times since regarding a wide array of environmental harms including air pollution and air pollution risk, Toxics Release Inventory facilities, noise pollution and more, some of the most important findings of this study include that

> Race proved to be the most significant among variables tested in association with the location of commercial hazardous waste facilities. This represented a consistent national pattern….
>
> Although socio-economic status appeared to play an important role in the location of commercial hazardous waste facilities, race still proved to be more significant. This remained true after the study controlled for urbanization and regional differences. [Though it is also true that] Incomes and home values were substantially lower when communities with commercial facilities were compared to communities in the surrounding counties without facilities (Commission for Racial Justice, 1987, p. xiii).

EJ and Health

One factor that is insufficiently stressed in the EJ literature is the important effects of pollution on health and the consequences of those effects. As stated flatly by Taylor (2014), "EJ research has to pay more attention to the relative danger of facilities" (p. 279). Health effects are, indeed, a primary concern of many analysts, especially those with degrees in health fields, such as Morello-Frosch. Wolch et al. (2014) says that "Community-based organizations, often aided by environmental groups, are refocusing urban brownfield remediation projects on urban green space to address public health and environmental justice concerns (Barnett, 2001)" (p. 239). But often the focus on health becomes blurred by the focus on methodology and the first-stage nearness and exposure that is easier to measure. And, indeed, the research presented in this book abstracts from actual health outcomes. But a powerful driver of our interest in EJ is that living in polluted environments has real health consequences, and we hope to retain focus on the importance of this fact.

In general, federally tracked pollutants are known human health hazards. As noted by its name, the Toxics Release Inventory (TRI), often used in EJ research (e.g., Campbell et al., 2010), tracks the release into the environment of products known to be *toxic* to humans. Other scholarship shows that air pollutants disproportionately impact minority communities (e.g., Morello-Froschet al., 2001; Kerr et al., 2021; Al-Kohlani et al., 2023).

> Common air pollutants with known health impacts were first regulated as 'Criteria Pollutants' by the 1970 Clean Air Act which established health-based National Ambient Air Quality Standards (NAAQS). The six criteria pollutants are carbon monoxide, ground-level ozone, lead, nitrogen dioxide, particulate matter, and sulfur dioxide" (Mecklenburg County Government, n.d.).

As overviewed in Campbell et al. (2023),

these pollutants cause a variety of physiological harms, and the cumulative effects of greater exposure to a whole variety of pollutants may partially explain race-based health disparities including worse birth outcomes for African Americans (Bekkar et al., 2020), greater death rates from COVID-19 for African Americans and Hispanics (Wu et al., 2020; Garcia Cano et al., 2020), and increased incidence of respiratory diseases such as asthma for African Americans and Latinx (Brulle & Pellow 2006; Brown, 1995; Mohai & Bryant, 1992; Szasz & Meuser, 1997).

Noise is another environmental harm that is often disproportionately imposed on minority communities. As reviewed in Sobotta et al. (2007), which finds Hispanic ethnicity to be the dominant predictor of which communities suffer the most airport noise, excess noise has been shown to cause a variety of health effects, from degraded hearing to high blood pressure to harm to fetuses. Olson (2018) provides evidence that, in San Francisco, schools in which a majority of students are members of minoritized groups are in noisier areas, and also that noise exerts an independent downward effect on test scores when holding constant other factors known to affect test scores.

Further, increasingly evidence shows that groups that are over-exposed to one tracked pollutant are likely to be overexposed to several (e.g., see information provided in the CalEnviroScreen, at https://oehha.ca.gov/calenviroscreen). Overall, degraded health reduces both the capacity to build human capital and also the ability to gain from the human capital one has developed (Gooden, 2015). The fact that racial and ethnic minorities are consistently, over decades and generations, disproportionately exposed to many environmental harms, may explain some of the health and performance gaps that we observe between, for example, Whites and African Americans in the US.

Most EJ literature focuses on the excess of environmental bads that impinge disproportionately on minoritized communities, but some newer research also focuses on deficits of environmental goods. Evidence is emerging that there is also race/ethnicity-based disproportionality regarding access to environmental amenities such as parks and greenspace (Wen et al., 2013; Gould & Lewis, 2017). Wolch, Byrne and Newell give an extensive overview of both direct and indirect health benefits of green amenities in urban environments (2014). And, while also noting that not all findings support this position, they provide evidence that the lack of green amenities is an EJ issue, too. For example, Wen et al. (2013) provides evidence that there is a negative association between proximity to, and size of, urban greenspace with poverty, percent African Americans, and percent Hispanics (Gould & Lewis, 2017, p. 13). Hsu et al. (2020) demonstrates that minoritized communities are more affected by the urban heat island (UHI) effect, and it is well known that UHI is reduced by greenspace. The Urban Heat Island effect can lead to excess heat, which can lead to heat stroke and respiratory and cardiovascular effects that can be fatal (CDC & APHA, 2018). Thus, minoritized groups are not only harmed by excess exposure to pollutants, but lack the healing effects of environmental amenities. Aside from just the benefit of cleaning up pollution, there is an additional and separate benefit from greening.

This subsection briefly introduces some history of EJ research and key issues of environmental injustice. Though there has certainly been consideration of other types of pollution such as air, noise, and water, EJ literature has often focused on land. As described above, land-use issues were the first issues, though they are not the only issues. But polluted land is especially salient partly because land is fixed in space. We start with a concern of simply cleaning up—health and justice concerns. Then we consider what happens if the cleaned up is also developed. And we also consider what happens if the cleaned up is greened up.

The next subsections address gentrification, displacement, and how these issues are related. As part of this we review the traditional gentrification literature, green gentrification literature, and then the just-green-enough literature, which is a recent addition that describes efforts by communities to realize some environmental improvement, but not enough improvement to appeal to potential gentrifiers.

Brief Review of Gentrification and Displacement

The original concept of gentrification arose in the 1960s in England and was explicitly class-based (Glass, 1964); in the US, racial and ethnic changes are also analyzed under this rubric. Now there are two related but different literatures, one the traditional gentrification literature, focused on market forces, economic development, and housing policy. The more-recent "green gentrification" approach is somewhat different; of course market forces and housing policy come into play, but the focus here is on explicit green-policy interventions—whether cleanup, addition of a new green amenity, or both—as gentrification triggers (Anguelovski et al., 2019; Anguelovski, et al., 2018). Thus, the focus on economic development as such is lessened in the green gentrification approach. In this subsection we briefly overview traditional gentrification concepts, and consider whether gentrification and displacement must go hand in hand. In the next subsection we discuss the newer concern with green gentrification.

Over the time that it has been studied as an urban phenomenon, gentrification has been viewed with different connotations: neighborhood improvement, or compositional shift and displacement. The former tends to focus on space, and the latter tends to focus on the people and/or the culture in neighborhoods. This definitional challenge reflects the duality of gentrification challenges for policymakers. On the one hand, central city communities in the United States, especially those comprised of poor and minoritized populations, are quite often lacking in investment, amenities, services, and job opportunities, so policymakers may be interested in ensuring that such investments flow to these communities; here the focus is often on the space, rather than the people. On the other hand, these investments often–but not always—foster a process whereby property values escalate rapidly owing to increased demand for homes in neighborhoods with new amenities, and this can displace the people for whom the investments were (putatively) intended; the concern here is on the people and the culture, rather than on the space. As noted by Richardson and colleagues, "Gentrification presents a challenge to communities that are trying to achieve economic revitalization without the disruption that comes with displacement" (2019, p. 4). Further, any displacement often takes on a class and racial tone, further complicating the effort to achieve investment and place improvement without displacing residents.

As mentioned above, the class-based tone was present in the origination of the term "gentrification." Glass (1964) originally used the term to describe "the influx of middle-class people displacing lower-class worker residents in urban neighborhoods" in London (Atkinson & Bridge, 2005, p. 4). Hyra (2016) also defines gentrification as "an influx of upper-income people to low-income areas" (p. 169). In these definitions, gentrification implies the population's "compositional shift" in urban neighborhoods: transitioning from one social group to another social group.

In US cities, compositional shifts toward White, prime-age, and college-educated households—not because of overall population change but due to the displacement of existing residents—are considered characteristics of neighborhood gentrifications occurring since 2000 (Hwang & Lin, 2016). For example, the disproportionate movement of educated Millennials to the central city in large municipalities makes them newcomers who enjoy gentrified neighborhoods but displace traditional residents who were often low-income minorities (Hyra, 2016).

From this perspective, gentrification includes displacement as part of the definition—that is, gentrification is about both the people and the space. However, the original focus on class is complicated in the United States: Though economic class remains a key consideration of gentrification, race also features prominently. Further, in the US context, it is quite clear that historic "economic development" such as "slum clearance" and freeway expansion often led to displacement of minoritized groups; in some cases it seems this was intentional (Taylor, 2014; Pulido, 2000).

In short, both people and space are intertwined with the concept of gentrification. This makes even the gentrification *conversation* difficult, let alone the difficulty of implementing policies and programs that aim to improve communities without completely changing them. The lack of conceptual clarity is evident in the varying definitions of gentrification in the literature, as depicted in Table 2.1, below. Different methodological traditions tend to conceive of gentrification differently, some including displacement as part of the process, and some viewing it as a separate (but related) process; the intentions behind the process are also viewed differently (Slater, 2006).

In sociologist Ruth Glass' (1964) original definition, displacement is perhaps the defining feature of gentrification and, not surprisingly, much of the work from sociologists continues this perspective. As mentioned in Chap. 1, urbanists often view gentrification more in the sense of a complex, "wicked" problem with multiple causes. Other definitions, such as Nelson's from 1988, focus more on the economic changes that take place, and while it may imply displacement, displacement is clearly a secondary or related consideration to a focus on the in-migration of wealthier residents. Many economically oriented studies use definitions more along these lines. Scholarship coming from action-oriented perspectives tend to see gentrification from the perspective of intentionality. For example, intentionality can be seen

Table 2.1 Conceptual definitions of gentrification

Author	Definition
Glass (1964, p. xix)	"Once this process of 'gentrification' starts in a district it goes on rapidly, until all or most of the original working-class occupiers are displaced and the whole social character of the district is changed"
Smith (1979, p. 139)	"[p]oor and working-class neighborhoods in the inner city are refurbished by an influx of private capital and middle-class homebuyers and renters"
Nelson (1988, p. 3)	"The process by which reinvestment in urban neighborhoods leads to an inflow of residents of higher socioeconomic status than the original residents of the community"
Hamnett (1991, p. 175)	"[g]entrification commonly involves the invasion by middle-class or higher-income groups of previously working-class neighbourhoods"
Kirkland (2008, p. 18)	"Gentrification is often a fundamentally racial transformation: The pre-gentrified neighborhood is inhabited mostly by African Americans or other people of color, and the in-movers are typically white."
Lees et al. (2010, p. xv)	"The transformation of a working-class or vacant area of the central city to a middle class residential and/or commercial use"
Hyra (2016, p. 169)	"Influx of upper-income people to low-income areas"
Merriam-Webster dictionary (retrieved 2024)	"a process in which a poor area (as of a city) experiences an influx of middle-class or wealthy people who renovate and rebuild homes and businesses and which often results in an increase in property values and the displacement of earlier, usually poorer residents"

in Hamnett's (1991) description of the process as an invasion, and in Smith's (1979) more benign, passive-voice description of "refurbishment." With its consideration of vacant areas, Lees et al. (2010) clearly allows for gentrification without displacement (on vacant land there is no one to displace), and Kirkland (2008) illustrates the American perspective that often focuses on racial change as central. Even dictionary definitions hedge on the connection between reinvestment and displacement (Merriam-Webster, 2024), as seen in Table 2.1 (note the modifier "often").

Knieriem (2023) suggests that part of the lack of conceptual clarity relates to the changing nature of gentrification itself, and that it is highly contextual to times and places, even while there are overlapping themes. They rely on a definition from Clark in 2010 that combines both aspects of investment in neighborhoods and displacement of original residents, but also note that in some cases race is a factor in addition to class (Fallon, 2021), and that some neighborhoods undergo super-gentrification (Lees, 2003) in which already gentrified neighborhoods move from upper-middle-class to wealthy, which seemingly removes race from consideration entirely. Easton et al. (2020) argue that the difficulty in empirically assessing displacement has, in part, fractured displacement from gentrification and provides an out for governments to focus policies on changes in space and ignore demographic changes in the people that made up a community.

While acknowledging the interconnectedness of both of these processes, in previous research we conceptually distinguish gentrification and displacement (Eckerd et al., 2019). In that research, we note that, although the processes are clearly related to one another, displacement does not necessarily follow from gentrification, and we sought to computationally consider different sets of circumstances in which displacement did or did not follow gentrification. We are not the only scholars to make this argument; Zuk et al. (2018) also argue that "Scholarship has generally conflated gentrification and displacement; however, this review argues for a clearer analytical distinction between the two" (p. 31). As reviewed in Chap. 1, Richardson et al. (2019) also provides empirical evidence of gentrification without displacement, and Gould and Lewis (2017) argue that in declining neighborhoods (such as Gary, Indiana, or Detroit) greening might induce neighborhood stability rather than cause gentrification.

However, as noted above, in the extreme case, "economic development" can explicitly and intentionally move people out of areas. As explained by Taylor (2014, p. 229),

> It is widely documented that urban renewal projects condemned and leveled many minority and poor communities that were labeled as 'blighted' or 'slums' to make way for commercial development, upscale residences, and public housing (Darden et al., 1987; Sugrue, 1996; Goodman & Monti, 1999; Gotham, 2001; Pritchett, 2003; Maskovsky, 2006; Blais, 2007).

In overviewing such cases, Taylor (2014) has a section specifically on eminent domain, which she argues has been used in these ways so often that it bears consideration as an EJ issue itself. In one sections she notes that "Fullilove (2007) studies 2532 eminent domain projects in 992 cities and found that between 1949 and 1973, these projects displaced a million people. Two-thirds of those who were displaced were Black" (Taylor, 2014, p. 233).

Jason Richardson, Bruce Mitchell, and Juan Franco, members of the National Community Reinvestment Coalition (NCRC), performed a study of gentrification and displacement at the level of the (normalized) US Census tract in 935 American metropolitan areas that covered a period from 2000 to 2013 (2019). This work used fairly strict definitions of both gentrification and displacement, and found high variability in both. They specifically focus on "residential gentrification and racial and ethnic displacement" (Richardson et al., 2019), considering that gentrification has occurred if a tract was gentrifiable (below 40th percentile for median income and housing value), and then there are increases in the college-educated and home values that are in the top 60th percentile, and real increases in median income. Once gentrification is identified, then to identify displacement they require two changes: a racial group's presence in a tract declines by "more than two standard deviations from the mean of all census tracts," and the "absolute number of residents from that racial group decline by at least [5%]." Gentrification was much more common in large and coastal cities, with a few large interior cities also experiencing gentrification. "Gentrification was centered on vibrant downtown business districts, and in about a quarter of the cases it was accompanied by racialized displacement" (Richardson et al., 2019, p. 5), specifically of Blacks and Hispanics. They found more displacement of Blacks in the South and more displacement of Hispanics in the West, and the numbers displaced can be quite significant, particularly for Blacks.

Though, as noted, it can't be observed in every gentrified census tract, within specific cities, 100s of persons *on average* from *every* tract with gentrification displacement can be displaced. For example, in Houston during the time studied, 8 gentrified census tracts showed evidence of Black displacement, with *on average* 648 Black persons being displaced from each of those 8 tracts (Richardson et al., 2019, p. 21, Fig. 4). In cities with high levels of Hispanic displacement during their study period, the average loss per tract in Denver and Austin was above 1000 persons (Richardson et al., p. 22). Given that census tracts have between 1200 and 8000 people "with an optimum size of 4000 people" (US Census Bureau, 2022), it seems quite likely that such losses damage social structures including social networks. As noted by many, those displaced are likely to be those with the fewest educational and monetary resources.

Most displacement research focuses on African Americans and Hispanics, but there is some research indicating that Asians may also suffer from environmental injustice (Campbell et al., 2010) and even displacement (Richardson et al., 2019, pp. 22, 28).[4]

Overview of Green Gentrification Literature

We are not the only ones to be concerned with "green gentrification"—gentrification and displacement that occurs because an area is environmentally improved. Though much newer than the traditional gentrification literature or the EJ literature, there is a growing literature on green gentrification (sometimes using other terms such as eco-gentrification). Activists such as those in Brooklyn's UPROSE group are focused on these concerns (Gould & Lewis, 2017), and so are an increasing number of scholars. Derickson et al. (2021) review several works that express such concern and find evidence of green displacement. Green gentrification has the potential to be especially a co-concern of brownfield cleanup—one of the key EJ issues—because land is immobile but people are not. If people were immobile, conversion of brownfield sites into something greener should clearly help nearby residents.

Since at least the 1980s there has been evidence that cleaning and greening can raise property values. For example, Taylor (2014) reports that

> Scholars studying this phenomenon have found that people *will pay more* for housing to avoid risks; they will also pay a premium once sites are cleaned up and disamenities are transformed into desired amenities. For example, Smith and Desvousges's (1986) study of suburban Boston found that residents were willing to pay more for housing in order to live further away from hazardous waste sites.
>
> Chattopadhyay et al. (2005) study of Waukegan, Illinois, found that residents were willing to pay more for housing when NPL [National Priority List, aka Superfund] sites around

[4] Though not highlighted by Richardson et al. (2019), their data do suggest that Asian displacement can also occur. See, e.g., p. 22 Table 5 and p. 28 Figure 9.

> Waukegan Harbor were cleaned up. Similarly Gayer et al. (2000, 2002) study of risk indicators in Grand Rapids, Michigan, found that residents were willing to pay more for housing to avoid perceived cancer risks from Superfund sites (pp. 79–80; italics in original).

However, attempts to reduce environmental injustice are not provided in isolation, but are embedded in cities' and communities' efforts to renovate and revitalize polluted, segregated, and low-income neighborhoods—and are also embedded in local power dynamics and urban growth-machine politics. Gould and Lewis (2017) explicitly argue that there is a newer type of green-growth-machine politics in which members of the urban growth machine now purposely use green changes to trigger growth. Trying to tackle environmental injustice without understanding neighborhood changes that result from pollution remediation and/or the addition of amenities can lead to unintended consequences that aggravate social injustice more broadly. Sometimes such changes are even intended,[5] and some may even use arguments around the de-concentration of poverty to argue that "displacement is exactly what" is needed (Tait, p. 18).

In standard economic analysis, pollution is an externality that leads to inefficiency. Within this frame, cleaning up a polluted site, such as a brownfield site, can be efficiency-enhancing. Similarly, adding greenspace provides positive externalities, and goods with positive externalities are generally undersupplied by market forces, so there is room for efficiency-enhancing government intervention. However, if such cleaning and greening engenders rising housing costs such as rent, then these are negative externalities to existing residents. While the positive benefits of cleaning and greening remain, there is also the question of who reaps the positive externalities and who bears the negative externalities.

Mixed Evidence of Green Displacement

It is, however, important also to note that not all studies have found green displacement. Gould and Lewis start by assuming that gentrification *does* displace, and this is reasonable given the original use of the term. But later they argue that *green* gentrification, anyway, doesn't always have to lead to displacement—on pages 15 and 23 they say displacement is not inevitable.

Taylor's careful analysis of cleanup literature also shows that cleanup does not always lead to minority displacement. For example, in research by Gamper-Rabindran and Timmins (2011) they use "restricted access fine-geographical-resolution block data" and find "Cleanup leads to increases in population density and housing unit density; increases in mean household income and shares of college-educated; but also to increases in the shares of minorities" (p. 620). In earlier

[5] An interesting example of this in Curran and Hamilton (2017) is one in which developers seemed to institute farmer's markets in order to cause gentrification and then abandoned them once achieved (see Curran & Hamilton, 2017, Chap. 6).

empirical work, Eckerd (2011) found that in the city of Portland, Oregon, widely considered to be an extra-green city, during the 1990s environmental improvements via hazardous site cleanups were not related to neighborhood-level gentrification. Importantly, some "studies suggest that there might be a period of time that neighborhoods *absorb* the gentrifiers without *displacement* of longtime low-income residents (Freeman, 2005; Freeman & Braconi, 2002a, b, 2004; Vigdor et al., 2002; McKinnish et al., 2008)" (Taylor, 2014, p. 88; italics in original).

In contrast, Essoka (2010) performed a pre-post study of brownfield sites: "The study first locates U.S. Environmental Protection Agency brownfields pilot sites residing in environmental justice communities. Second, a pre-post demographic study evaluates gentrification variables at the selected locations. Finally, statistical analyses determine whether environmental justice populations experience displacement" (p. 299). Essoka (2010) found evidence of important displacement after brownfield cleanup and redevelopment: "The study found that Blacks were displaced 61% of the time and Latinos 14% of the time in brownfield redevelopment projects" (as quoted in Taylor, 2014, p. 88). In their book entitled *Green Gentrification*, Gould and Lewis (2017) carefully examine 5 cases in the Brooklyn borough of New York City (NYC), and find clear evidence of green displacement in 4 of the 5 cases (though in one case the neighborhood was already "super-gentrified" so the displacees are very different—"the 1 percent is pushing out the 5 percent" (p. 79)), with the outcome for the fifth still unclear as of publication. One of the cases they analyze is a case of restoration of an existing environmental amenity (Prospect Park) but the others are cases of brownfield cleanup. In one case (Gowanus Canal), though there is clear evidence of gentrification, there is also a real question whether anyone should live there as "greening" is a relative concept: It does not yet seem clean enough to be safe!

In some ways, NYC is unique. As pointed out by Gould and Lewis (2017) themselves, NYC may be a special case since it is a global city with significant international finance and international real-estate investment, and also a city in which there is strong evidence of continuing and explicit racial discrimination in housing (p. 11). After completion of their analyses, Gould and Lewis infer that the "globalness" of the city and the "hotness" of real estate matters (2017, pp. 164–165). But Gould and Lewis also state that "Urban greening is a global phenomenon evident in cities around the world from New York to Cape Town to Shanghai" (p. 151). Further, Immergluck and Balan (2018) find similar results for Atlanta's Beltline, though Atlanta is clearly not the same type of international finance and real-estate hub. Wolch et al. (2014) says there is evidence in China that "Even the smallest green space embellishments may drive up property prices in the urban core, where densities are highest, parks are fewer, and temperatures are the hottest" (p. 238).

Overall, the wide variation in findings regarding green gentrification could be because local conditions are quite different, and both the scale and type of cleanup vary from place to place. "The urban form, or patterns of land-use in US cities, differ considerably in size and structure, and the process of gentrification can involve neighborhoods in a rapid process of change or unfold over decades in larger districts" (Richardson et al., 2019, pp. 7–8). As they are wrapping up their book, Gould and Lewis (2017) argue that people often ignore the unique environments of greened cities such as Curitiba (p. 165).

One of our points is to avoid this error; we do not want to analyze as if the same solutions are likely to work everywhere. As we explain in other chapters, the differences in findings overviewed above suggest it may be the configural conditions that matter in the effects of cleaning and greening policy interventions.

Proposed Solutions

With all the concern about gentrification and displacement, green and otherwise, it is natural to ask what solutions are proposed. It is in the nature of cities to be in constant states of change, and areas that see gentrification with displacement usually needed improvement, so stopping neighborhood change is not our goal. How might policy guide change in desirable ways? Gould and Lewis (2017) argue that "calls for urban greening without displacement are calls for non-market approaches to sustainability" (pp. 158–159). Pearsall and Anguelovski (2016, p. 4) reports that "In Malmö, Sweden, Fitzgerald and Lenhart (2016) found that Eco-Districts were not associated with gentrification because city planners had an explicit goal to include low-to-moderate income housing in all districts." Review of suggestions in the literature makes it clear that scholars do not expect that markets alone will solve the displacement problem, because there is a recognition that market forces are a driver of the problem. Our focus on policy, rather than markets, accords with this. We do not expect that pure market forces will prevent displacement; instead, intentional public policy must be brought to bear.

In a paper that reviews the gentrification literature from an urban-planning perspective, Tait (2015) focuses on "blight" and argues that higher-income gentrification displacement is the standard outcome from remediating urban blight, but that there are four strategies in the urban planning literature that are viewed as promising for reducing displacement: "the gentrification with justice model, asset based community development, urban renewal, and adaptive reuse of industrial brownfields" (p. 1). Tait (2015) defines "a successful blight remediation effort as one that increases property values and increases population while minimizing the displacement of indigenous residents or the loss of historic character"[6] (p. 2). From the perspective of the green gentrification literature it is interesting that adaptive reuse of brownfields is viewed as promising in the urban-planning literature, for this is shown to trigger green gentrification and displacement in many cases (some reviewed above) and is one of the key concerns of green gentrification. From the cases that Tait examines, it may be that the promising urban planning solutions are most possible

[6] Note that Tait's definition of a brownfield is non-standard. In the EJ focus the definition emphasizes known or suspected environmental contamination, but Tait focuses on abandoned "commercial or industrial properties that are above average in square footage" (p. 6). Also note that our own analyses ignore historic character.

when there is significant vacancy and underuse of land and property in a neighborhood—and indeed that is part of Tait's definition of "blight" (2015, p. 4), and his definition of urban redevelopment relies heavily on this assumption. This is one indicator that density may be an important variable.

In their discussion of potential solutions, Richardson and his NCRC colleagues include community-group participation including "partnerships between banks and community-based organizations to encourage equitable development," and a variety of affordable-housing-focused solutions including "limited-equity co-ops and community land trusts; providing existing tenants with the right of first refusal in apartment conversions coupled with low-income and first-time buyer financing programs…and split tax rates" (Richardson et al., 2019, p. 6). Similarly, Curran and Hamilton (2017, p. 5) report that Faber and Kimelberg (2014) make a distinction between market-based and institutional approaches ("focus on access to housing"), activist approaches ("challenging narratives of what green should look like"), and contractual approaches ("focus on community benefits guarantees").

Fundamentally, the solutions discussed above assume that sufficient gentrification for displacement will occur, and then try to stop displacement through a variety of means. An overarching proposal that is gaining traction is the just green enough approach, discussed in the next section. This approach is different in that it tries to improve neighborhoods such that they gentrify only to a level that will *not* cause displacement.

Just Green Enough

Some scholars and community activists argue that one way to prevent green gentrification is to make places just green enough (JGE) (see, e.g., Wolch et al., 2014; Curran & Hamilton, 2017). The balance here is a little tricky. The idea is to engage in cleaning and greening that improves the lot for those already living in a polluted neighborhood, without providing an environment that is so much cleaner and/or greener as to foster a gentrification cascade leading to displacement. The JGE approach can make one worried that proponents are saying that polluted minority neighborhoods deserve cleaner and greener environments—but not to the level of better-off Whiter neighborhoods. However, one important issue is what residents prefer: They may be willing to trade off some amenities for a reduced probability of displacement. Another acceptable way disadvantaged neighborhoods might benefit from a JGE approach is if those living in an area have at least somewhat different taste regarding amenities than preferences held by better-off potential gentrifiers.

In their edited book, *Just Green Enough*, Curran and Hamilton state,

> just green enough was meant to describe a plan for remediation and revitalization that uncouples environmental cleanup from high-end residential and commercial development,…. [T]he activism that gave rise to the just-green-enough approach was focused explicitly on social justice and environmental goals *as defined by the local community*, those people who have been most negatively affected by environmental disamenities, with

> the goal of keeping them in place to enjoy any environmental improvements (2017, p. 3; italics added).

Throughout the edited book Curran and Hamilton show awareness of the careful balancing act needed to take a JGE stance. Later in the work they explicate: "So, ideally, cleanup…will be just green enough to improve the health and quality of life of existing residents, but not so literally green as to attract upscale…developments that drive out working class residents…." (p. 16).[7] "For JGE advocates, the goal has always been to reduce environmental health risk, not to accept more" (Curran & Hamilton, 2017, p. 34).

As pointed out by Walker (2012), one form of environmental justice focuses on procedural justice. Discussing greenspace, he argues the following:

> This brings us to another perspective, which is to frame justice and greenspace as a matter of participation and procedural justice. There are many grounds for involving local communities in the process of planning, investing in, establishing and maintaining public greenspaces. There are substantive arguments that through participation greenspaces will be better designed, better reflect local needs and values, better address the needs of multiple users and also tap into local creativity and enthusiasms (p. 176).

Such ideas of involving the pre-gentrification community may help with developing JGE areas—areas that appeal to the locals rather than to potential higher-income and Whiter gentrifiers.

The question then becomes what *is* just green enough? How do we help local decisionmakers go green enough and not too green? *Just Green Enough* provides case studies, many within NYC (including Brooklyn), but also including Vancouver, San Diego, Seoul, Chennai, Chicago, Los Angeles, Philadelphia, and more—but still leaves us to wonder: What is generalizable from these cases? As noted by Curran and Hamilton (2017, p. 8), the goal is "effectively …creating change within different social, economic, historical, and policy contexts." These authors, and others, including Agyeman (2013), call for new policy and planning approaches (p. 10), and we hope to help along this route. Curran and Hamilton state: "city policy can make a difference" (2017, p. 33). But what policy is not yet known. We hope that our analyses in this book will help break some paths.

One question that comes up from a JGE approach is the size and scale of cleaning and greening; might a small park be just green enough, while a large one leads to displacement? During the periods studied by Gould and Lewis (2017) and Curran and Hamilton (2017), a very interesting policy intervention in NYC was that PlaNYC had a goal of all New Yorkers living within a 10-minute walk of a park.

[7] In this work our focus is on residents, not on firms. But it is worth noting that one concern of many scholars around excessive greening is that it drives out the often-industrial jobs that existing residents use as their employment base (e.g., Curran & Hamilton, 2017, p. 16 and elsewhere). Many argue the JGE approach also needs to maintain working-class jobs (e.g., Curran & Hamilton p. 28 and other places; UPROSE activist community; see e.g., https://www.uprose.org/campaigns). But Curran and Hamilton (2017, p. 33) find that "The maintenance of industrial zoning, a central element to the just green enough strategy, is not, in and of itself, enough to prevent gentrification."

> This inspired 'New York's third great era of park building' (The City of New York, 2007), [and] a major initiative was to 'target high-impact projects in neighborhoods underserved by parks' (The City of New York, 2007)… a quarter million of the two million New Yorkers who had not had access, gained ten-minute walking access through those efforts (Green NYC, 2016) (Gould & Lewis, 2017, p. 64).

As presented by Gould and Lewis, there is no indication that this intervention led to gentrification displacement, which provides a glimmer of hope that scale may be relevant to successful anti-displacement policies. In other cases, Pearsall and Anguelovski (2016, p. 4) report that

> the design and scale of green space may matter. Small-scale environmental projects may have less potential to attract large-scale development (Wolch et al., 2014), and long linear green spaces integrated with blue infrastructure and with affordable public transit systems may make access to green space more equitable over time and space (Ngom et al., 2016)".[8]

Thus, one factor we examine in Chaps. 6 and 7 is scale of green intervention.

What Are We Doing Differently?

Many of the scholars interested in gentrification and displacement are sociologists and social geographers. Their work focuses significantly on class distinction; it is rooted solidly in the initial conception of "gentry-fication" developed by Glass (1964), herself a sociologist. Such scholars are importantly concerned with *income-class* winners and losers of neighborhood change.

We are policy analysts whose concern with gentrification and displacement grows from an EJ perspective. That race alone should matter to observable environmental exposure is clearly a violation of efficiency expectations, but evidence from 40 years of EJ research shows that it often does. Because of our explicitly EJ focus, we are most concerned with disproportionately polluted *race/ethnicity* winners and losers of cleaning and greening policies. Thus, we are specifically focused on disproportionate environmental harm to minority populations, the goal of improved social justice through cleanup and greenup, and how policy-makers might predict what is green enough.

One way we are in full accord with other scholars of green gentrification is that we, too, argue that those who bear the external costs of pollution in their neighborhoods should receive the external benefits of cleanup. These benefits should not go to gentrifiers, whether developers or incomers, *instead of* existing residents. The invisible hand problem is that improving an area in a thriving city makes that land more valuable, and that can set into play strong market forces that push out those who social justice says should benefit. According to Dooling (2018), especially during the Reagan and Thatcher eras, "The process of upending working class and low-income neighborhoods was perceived as the unavoidable outcome of the

[8] "Blue infrastructure" indicates water elements, such as rivers, etc.

invisible hand underlying urbanization…." (p. 47). But, as Dooling points out later, "developers are often granted tax breaks and awarded other subsidies that city politicians justify by claiming the economic prosperity of the city is dependent upon tax revenue" (Dooling, 2018, p. 47). This raises a broader point: As they exist today, large cities are by no means run by an unregulated invisible hand! Neighborhoods, citizens, firms, pollution, greenspace, cleanup—all are heavily regulated. Moreover, the necessity of regulation in cities has been recognized for more than 100 years due to the many externalities of living densely. Thus, the question is what regulation is beneficial, not whether to have it or not. Further, to a significant extent, the benefits of cleaning and greening are created by society via collective government, and the benefits shouldn't just be scooped up by developers and the better-off; they should be distributed in ways that also consider social equity.

A simple view of Cost-Benefit Analysis (CBA) might suggest that developers and the better off should receive the benefits of cleaning and greening because they have more willingness-to-pay (WTP) for these benefits. In this view, then, gentrification and displacement would be efficient urban changes. But CBA is meant to be Pareto-improving and, in its pure form, requires not only that benefits exceed costs, but that the outcome be Pareto-improving—so there are no losers. We often forget that we use a second-order condition, the Kaldor-Hicks principle,[9] on the assumption that, on average, people who lose from one policy gain from another, and that, since society is net better off, then differences between groups are not systematic and wash out. Unfortunately, significant evidence indicates that this is not true regarding minorities and the city. As noted by Pulido (2000) and many others, disproportionate negative outcomes for minoritized groups have become systemic in US cities, and these groups lose again and again. In such cases, greater WTP is an insufficient justification for scooping social benefits.

Other authors also do consider policy issues, but they are generally less focused on policy than we. For example, Gould and Lewis (2017) discuss zoning issues multiple times (e.g., p. 142, 144), and of course zoning is one important type of urban policy. But they put most emphasis on social movements (e.g., p. 36, 126)—which is not policy. On the other hand, Gould and Lewis state that "The just green enough approach to urban sustainability seeks to identify that tipping point in the urban environmental amenity production process where an area becomes attractive to growth coalitions…. and that boundary is likely to be largely defined by how hot or cold a real estate market is" (p. 170). They go on to say that where that is will vary by city and by neighborhood, stating that "there is a wide range of neighborhoods with very different green-gentrification tipping points" (p. 171). Note that this approach focuses on just one variable—and a market-based one at that. We very much agree with their points regarding variation: We agree that the tipping point will vary by city and by neighborhood conditions. But, as we argue elsewhere in this book, our approach is one that views the city as a complex system, arguing that

[9] The Kaldor-Hicks principle asserts that we can adopt a policy if the benefits exceed the costs such that the winners could benefit the losers. For actual Pareto efficiency, this compensation must occur—a point that is often forgotten

focusing on any one (or two) variables—be they zoning or real-estate markets—is insufficient: it is configural clusters of variables that are needed to help forecast the effects of cleaning and greening.

How can policymakers understand where and how much they can clean and green without driving out those who deserve to benefit to receive partial recompense for the harm they have endured? In order to gain leverage on this question, we believe we must move beyond case studies, still using information from them, but also using insights from complexity theory to bring forth general guidance for cleaning and greening policy in the service of social justice. The next chapter focuses on this complexity perspective and its implications for our study.

References

Agyeman, J. (2013). *Introducing just sustainabilities: Policy, planning, and practice*. Zed Books.

Al-Kohlani, S. A., Campbell, H. E., & El-Khatib, S. O. (2023). Minority faith and environmental justice. *Politics & Policy*.

Anguelovski, I., Connolly, J. J. T., Masip, L., & Pearsall, H. (2018). Assessing green gentrification in historically disenfranchised neighborhoods: A longitudinal and spatial analysis of Barcelona. *Urban Geography, 39*(3), 458–491.

Anguelovski, I., Connolly, J. J., Garcia-Lamarca, M., Cole, H., & Pearsall, H. (2019). New scholarly pathways on green gentrification: What does the urban 'green turn' mean and where is it going? *Progress in Human Geography, 43*(6), 1064–1086. https://doi.org/10.1177/0309132518803799

Atkinson, R., & Bridge, G. (2005). *Gentrification in a global context*. Routledge.

Barnett, H. (2001). The Chinatown cornfields: Including environmental benefits in environmental justice struggles. *Critical Planning, 8*, 50–60.

Bekkar, B., Pacheco, S., Basu, R., & DeNicola, N. (2020). Association of air pollution and heat exposure with preterm birth, low birth weight, and stillbirth in the US: A systematic review. *JAMA Network Open, 3*(6), e208243. https://doi.org/10.1001/jamanetworkopen.2020.8243

Bell, D. (2004). Environmental justice and Rawls' difference principle. *Environmental Ethics, 26*(3), 287–306. https://doi.org/10.5840/enviroethics200426317. As cited in Walker (2012).

Blais, L. E. (2007). Urban revitalization in the Post-Kelo Era. *Fordham Urban Law Journal, 34*(2), 657–688. As cited in Taylor (2014).

Brown, P. (1995). Race, class, and environmental health: A review and systematization of the literature. *Environmental Research, 69*(1), 15–30. https://doi.org/10.1006/enrs.1995.1021

Brulle, R. J., & Pellow, D. N. (2006). Environmental justice: Human health and environmental inequalities. *Annual Review of Public Health, 27*(1), 103–124. https://doi.org/10.1146/annurev.publhealth.27.021405.102124

Bullard, R. D. (1983). Solid waste and the Black Houston Community. *Sociological Inquiry, 53*(2–3, 273), –288. https://doi.org/10.1111/j.1475-682X.1983.tb00037.x

Campbell, H. E., Peck, L. R., & Tschudi, M. K. (2010). Justice for all? A cross-time analysis of toxics release inventory facility location. *Review of Policy Research, 27*(1), 1–25.

Campbell, H. E., Kim, Y., & Eckerd, A. (2015). *Rethinking environmental justice in sustainable cities: Insights from agent-based modeling*. Routledge.

Campbell, H. E., Kim, S., Johnson, S., & Cáceres, C. (2023). Environmental racism and air pollution: Pre and post the COVID-19 economic shutdown. *Review of Policy Research*. https://doi.org/10.1111/ropr.12570

Capps, K. (2014, July 29). How much cleaning up brownfields is really worth: A new study shows the enormous effect that the EPA's brownfield remediation program has on real estate values in cities. *Bloomberg*. https://www.bloomberg.com/news/articles/2014-07-29/how-much-cleaning-up-brownfields-is-really-worth

Centers for Disease Control and Protection & American Public Health Association. (2018). *Extreme heat can impact your health in many ways*. https://www.cdc.gov/climateandhealth/pubs/extreme-heat-final_508.pdf

Chattopadhyay, S., Braden, J. B., & Patunru, A. (2005). Benefits of hazardous waste cleanup: New evidence from survey- and market-based property value approaches. *Contemporary Economic Policy, 23*(3), 357–375. https://doi.org/10.1093/cep/byi027 As sited in Taylor (2014).

Clark, E. (2010). The order and simplicity of gentrification–A political challenge. In L. Lees, T. Slater, & E. Wyly (Eds.), *The gentrification reader* (pp. 24–30). Routledge.

Commission for Racial Justice. (1987). *Toxic wastes and race in the United States: A national report on the racial and socio-economic characteristics of communities with hazardous waste sites*. United Church of Christ.

Curran, W., & Hamilton, T. (Eds.). (2017). *Just green enough: Urban development and environmental gentrification*. Routledge.

Darden, J. T., Hill, R. C., Thomas, J, & Thomas, R. (1987). *Detroit: Race and uneven development*. Temple University Press. As cited in Taylor (2014).

Derickson, K., Klein, M., & Keeler, B. L. (2021). Reflections on crafting a policy toolkit for equitable green infrastructure. *Npj Urban Sustainability, 1*(21). https://doi.org/10.1038/s42949-021-00014-0

Dooling, S. (2018). Making just green enough advocacy resilient: Diverse economies, ecosystem engineers and livelihood strategies for low-carbon futures. In W. Curran & T. Hamilton (Eds.), *Just green enough* (pp. 47–60). Routledge.

Easton, S., Lees, L., Hubbard, P., & Tate, N. (2020). Measuring and mapping displacement: The problem of quantification in the battle against gentrification. *Urban Studies, 57*(2), 286–306.

Eckerd, A. (2011). Cleaning up without clearing out? A spatial assessment of environmental gentrification. *Urban Affairs Review, 47*(1), 31–59. https://doi.org/10.1177/1078087410379720

Eckerd, A., Kim, Y., & Campbell, H. (2019). Gentrification and displacement: Modeling complex urban process. *Housing Policy Debate, 29*(2), 273–295. https://doi.org/10.1080/10511482.2018.1512512

Essoka, J. D. (2010). The gentrifying effects of brownfields redevelopment. *Western Journal of Black Studies, 34*(3), 299–315.

Faber, D., & Kimelberg, S. (2014). Sustainable urban development and environmental gentrification: The paradox confronting the US environmental justice movement. In H. R. Hall, C. C. Robinson, & A. Kohli (Eds.), *Uprooting urban America: Multidisciplinary perspectives on race, class & gentrification*. Peter Lang. As cited in Curran and Hamilton (2018).

Fallon, K. F. (2021). Reproducing race in the gentrifying city: A critical analysis of race in gentrification scholarship. *Journal of Race, Ethnicity and the City, 2*(1), 1–28.

Fillilove, M. (2007). *Eminent domain and African Americans: What is the price of the commons?* Institute for Justice. February. As cited in Taylor 2014.

Fitzgerald, J., & Lenhart, J. (2016). Eco-districts: Can they accelerate urban climate planning? *Environment and Planning C: Government and Policy, 34*(2), 364–380. https://doi.org/10.1177/0263774X15614666

Freeman, L. (2005). Displacement or succession? Residential mobility in gentrifying neighborhoods. *Urban Affairs Review, 40*(4), 463–491. As cited in Taylor 2014.

Freeman, L. (2006). *There goes the 'hood: Views of gentrification from the ground up*. Temple University Press. As cited in Taylor (2014).

Freeman, L., & Braconi, F. (2002a). Gentrification and displacement. *Urban Prospects, 8*(1), 1–4. As cited in Taylor 2014.

Freeman, L., & Braconi, F. (2002b). *Gentrification and displacement: New York City in the 1990s* [Unpublished manuscript]. As cited in Taylor 2014.

Freeman, L., & Braconi, F. (2004). Gentrification and displacement in New York City. *Journal of the American Planning Association, 70*(1), 39–52. As cited in Taylor 2014.

Frost, R. (2020, May). *Are Americans stuck in place? Declining residential mobility in the US*. Joint Center for Housing Studies of Harvard University. https://www.jchs.harvard.edu/sites/default/files/harvard_jchs_are_americans_stuck_in_place_frost_2020.pdf

Gamper-Rabindran, S., & Timmins, C. (2011). Hazardous waste cleanup, neighborhood gentrification, and environmental justice: Evidence from restricted access census block data. *American Economic Review, 101*(3), 620–624.

Garcia Cano, R., Snow, A., & Anderson, B. (2020, June 19). Covid-19 is ravaging America's vulnerable Latino communities. *AP News*. https://apnews.com/article/honduras-az-statewire-race-and-ethnicity-suburbs-virus-outbreak-d3c55be27a2549afa2906c190fe5adc9

Gayer, T., Hamilton, J. T., & Viscusi, W. K. (2000). Private values of risk tradeoffs at superfund sites: Housing market evidence on learning about risk. *Review of Economics and Statistics, 82*(3), 439–451. As sited in Taylor (2014).

Gayer, T., Hamilton, J. T., & Viscusi, W. K. (2002). The market value of reducing cancer risk: Hedonic housing prices with changing information. *Southern Economic Journal*, *69*(2), 266–289. As sited in Taylor (2014).

Glass, R. (1964). *London: Aspects of change*. MacGibbon and Kee.

Gooden, S. T. (2015). *Race and social equity: A nervous area of government*. Routledge.

Goodman, M. D., & Monti, D. J. (1999). Corporately sponsored redevelopment campaigns and the social stability of urban neighborhoods: St. Louis revisited. *Journal of Urban Affairs, 21*(1), 101–127. As cited in Taylor (2014).

Gotham, K. F. (2001). A city without slums: Urban renewal, public housing, and downtown revitalization in Kansas City, Missouri. *American Journal of Economics and Sociology, 60*(1), 285–316.

Gould, K. A., & Lewis, T. L. (2017). *Green gentrification: Urban sustainability and the struggle for environmental justice*. Routledge.

Green, N. Y. C. (2016). Mike *Bloomberg's Record of Progress*. Retrieved January 20, 2016 at http://progress.mikebloomberg.com/green-nyc/parks

Hamnett, C. (1991). The blind men and the elephant: The explanation of gentrification. *Transactions of the Institute of British Geographers*, 173–189.

Hsu, A., Sheriff, G., Chakraborty, T. C., & Manya, D. (2020). *Disproportionate exposure to urban heat island across major US cities*. Available at SSRN 3684952 or https://doi.org/10.2139/ssrn.3684952.

Hwang, J., & Lin, J. (2016). What have we learned about the causes of recent gentrification? *City, 18*(3), 9–26. https://doi.org/http://www.jstor.org/stable/26328271

Hyra, D. (2016). Gentrification commentary: Causes, consequences, and the future of equitable development policy. *City, 18*(3), 1–11.

Immergluck, D., & Balan, T. (2018). Sustainable for whom? Green urban development, environmental gentrification, and the Atlanta Beltline. *Urban Geography, 39*(4), 546–562. https://doi.org/10.1080/02723638.2017.1360041

Kerr, G. H., Goldberg, D. L., & Anenberg, S. C. (2021). COVID-19 pandemic reveals persistent disparities in nitrogen dioxide pollution. *Proceedings of the National Academy of Sciences, 118*(30), 1–8.

Kirkland, E. (2008). What's race got to do with it? Looking for the racial dimensions of gentrification. *Western Journal of Black Studies, 32*(2), 18–30.

Knieriem, M. (2023). Why can't we grasp gentrification? Or: Gentrification as a moving target. *Progress in Human Geography, 47*(1), 3–23.

Koo, C. (2021). *How social mix brings about social outcomes: Mechanisms and relevant factors* (Order No. 28868649). Available from Dissertations & Theses @ Claremont University Consortium; ProQuest Dissertations & Theses Global; Social Science Premium Collection. (2621218596). http://ccl.idm.oclc.org/login?url=https://www.proquest.com/dissertations-theses/howsocial-mix-brings-about-outcomes-mechanisms/docview/2621218596/se-2

Lee, S. (2023, May). *Tax revolt and neighborhood change: Effect of assessment limits* [Paper presentation]. 2023 Urban Economics and Public Finance Conference.
Lees, L. (2003). Super-gentrification: The case of Brooklyn Heights, New York City. *Urban Studies, 40*(12), 2487–2509.
Lees, L., Slater, T., & Wyly, E. K. (Eds.). (2010). *The gentrification reader* (Vol. 1). Routledge.
Levine, D., & Aharon-Gutman, M. (2023). There's no place like real estate: The "self-gentrification" of homeowners in disadvantaged neighborhoods facing urban regeneration. *Journal of Housing and the Built Environment, 38*(2), 775–794. https://doi.org/10.1007/s10901-022-09970-0
Maldonado, M. (1996). Brownfields boom. *Civil Engineering, 66*(5), 36. As cited in Tait (2015).
Maskovsky, J. (2006). Governing the "new hometowns": Race, power, and neighborhood participation in the new inner city. *Identities: Global Studies in Culture and Power, 13*(1), 73–99. As cited in Taylor (2014).
McKinnish, T., Walsh, R., & White, T. K. (2008). *Who gentrifies low-income neighborhoods?* [Working Paper]. National Bureau of Economic Research. As cited in Taylor (2014). https://www.nber.org/system/files/working_papers/w14036/w14036.pdf
Mecklenburg County Government. (n.d.). *Criteria air pollutants.* https://www.mecknc.gov/LUESA/AirQuality/EducationandOutreach/Pages/CriteriaPollutants.aspx
Merriam-Webster. (2024). Gentrification. In *Merriam-Webster.com dictionary*. Retrieved May 9, 2024, from https://www.merriam-webster.com/dictionary/gentrification
Mohai, P., & Bryant, B. (1992). Environmental racism: Reviewing the evidence. In *Race and the incidence of environmental hazards: A time for discourse* (pp. 163–176). Westview.
Morello-Frosch, R., Pastor, M., & Sadd, J. (2001). Environmental justice and Southern California's "riskscape:" The distribution of air toxics exposures and health risks among diverse communities. *Urban Affairs Review, 36*(4), 551–578. https://doi.org/10.1177/10780870122184993
National Environmental Justice Advisory Council. (2006). *Unintended impacts of redevelopment and revitalization efforts in five environmental justice communities.* National Environmental Justice Advisory Council. August. As cited in Gould and Lewis (2017).
Nelson, K. (1988). *Gentrification and distressed cities.* University of Wisconsin Press.
Ngom, R., Gosselin, P., & Blais, C. (2016). Reduction of disparities in access to green spaces: Their geographic insertion and recreational functions matter. *Applied Geography, 66*, 35–51.
Olson, E. (2018). *Education policy and environmental justice: Noise pollution, education performance indicators and the perpetuation of socioeconomic status* (Publication No. cgu11196). [Doctoral dissertation, Claremont Graduate University]. ProQuest Dissertations & Theses Global.
Organisation for Economic Co-operation and Development. (2021, May 31). *Brick by brick: Building better housing policies*. OECD. https://doi.org/10.1787/b453b043-en.
Pearsall, H., & Anguelovski, I. (2016). Contesting and resisting environmental gentrification: Responses to new paradoxes and challenges for urban environmental justice. *Sociological Research Online, 21*(3), 121–127.
Pritchett, W. E. (2003). The "public menace" of blight: Urban renewal and the private uses of eminent domain. *Yale Law & Policy Review, 21*, 1–52. As cited in Taylor (2014).
Pulido, L. (2000). Rethinking environmental racism: White privilege and urban development in southern California. *Annals of the Association of American Geographers, 90*(1), 12–40.
Richardson, J., Mitchell, B., & Franco, J. (2019, March 19). *Shifting neighborhoods: Gentrification and cultural displacement in American cities.* NCRC: National Community Reinvestment Coalition. https://ncrc.org/gentrification/
Slater, T. (2006). The eviction of critical perspectives from gentrification research. *International Journal of Urban and Regional Research, 30*(4), 737–757.
Smith, N. (1979). Toward a theory of gentrification: A back to the city movement by capital not people. *Journal of the American Planning Association, 45*, 538–548.
Smith, N. (1996). *The new urban frontier: Gentrification and the revanchist city*. Routledge. As cited in Curran and Hamilton (2018).

Smith, V. K., & Desvousges, W. H. (1986). The value of avoiding a LULU: Hazardous waste disposal sites. *Review of Economics and Statistics*, *68*, 293–299. As sited in Taylor (2014).
Sobotta, R. R., Campbell, H. E., & Owens, B. J. (2007). Aviation noise and environmental justice: The barrio barrier. *Journal of Regional Science, 47*(1), 125–154.
Sugrue, T. J. (1996). *The origins of the urban crisis: Race and inequality in postwar Detroit*. Princeton University Press. As cited in Taylor (2014).
Szasz, A., & Meuser, M. (1997). Environmental inequalities: Literature review and proposals for new directions in research and theory. *Current Sociology, 45*(3), 99–120.
Tait, A. (2015). *Multidimensional urban development schemes: The challenging task of remediating blight while minimizing residential displacement*. Paper for Urban Studies Senior Seminar, Rhodes College.
Taylor, D. E. (2014). *Toxic communities: Environmental racism, industrial pollution, and residential mobility*. New York University Press.
The City of New York (2007). *PlaNYC: A greener greater New York*. The City of New York. Mayor Michael R. Bloomberg. New York: NYC Government.
US Census Bureau. (2022, April 11). *Glossary*. https://www.census.gov/programs-surveys/geography/about/glossary.html
Vigdor, J. L., Massey, D. S., & Rivlin, A. M. (2002). Does gentrification harm the poor? [With Comments]. *Brookings-Wharton Papers on Urban Affairs*, 133–182. As cited in Taylor 2014.
Walker, G. (2012). *Environmental justice: Concepts, evidence and politics*. Routledge.
Wen, M., Zhang, X., Harris, C. D., Holt, J. B., & Croft, J. B. (2013). Spatial disparities in the distribution of parks and green spaces in the USA. *Annals of Behavioral Medicine, 45*(1), S18–S27.
Wolch, J. R., Byrne, J., & Newell, J. P. (2014). Urban green space, public health, and environmental justice: The challenge of making cities 'just green enough'. *Landscape and Urban Planning, 125*, 234–244.
Wu, X., Nethery, R. C., Sabath, M. B., Braun, D., & Dominici, F. (2020). Exposure to air pollution and covid-19 mortality in the United States: A nationwide cross-sectional study. *Medrxiv : The Preprint Server for Health Sciences, 2020*. https://doi.org/10.1101/2020.04.05.20054502.
Zuk, M., Bierbaum, A. H., Chapple, K., Gorska, K., & Loukaitou-Sideris, A. (2018). Gentrification, displacement, and the role of public investment. *Journal of Planning Literature, 33*(1), 31–44. https://doi.org/10.1177/0885412217716439

Chapter 3
Complexity Approach to Urban Systems

Abstract This chapter describes the view that cities are complex adaptive systems (CAS) and discusses some implications of this view, including that simple individual behaviors lead to unanticipated, emergent, system-wide outcomes; that configurations of urban contexts matter when striving to understand the effects of a policy lever; and that urban problems may be "wicked." We then introduce two research approaches suitable for studying complex urban systems, Qualitative Comparative Analysis (QCA) and Agent-Based Modeling (ABM). The chapter concludes with a discussion of the implications of the knowledge produced by these perspectives on cities and research methods.

Introduction

Given our view that the problems of environmental justice (EJ) and gentrification are intertwined in the social systems of cities, we view urban policy decision-making through the lens of complexity and Complex Adaptive Systems (CAS). From this perspective, we seek to understand how environmental improvements interact with social systems to produce unintended consequences, like displacement, and, importantly, to consider how policy implemented into these complex environments can consider the nature of interdependent systems. Each chapter in this book looks at this challenge. In this chapter, we focus on explaining the perspective that cities are CAS, and that this reality is central to understanding the effects of urban policy interventions.

Many of the systems we experience as a part of our lives are a collection of diverse and complex processes that phenomenal beings create on a daily basis. Simply put, a system is a set of elements—organizations, people, cells, or whatever—interconnected in such a way that they can produce their own pattern of behavior over time (Meadows, 2008). CAS is one in which many independent elements or agents interact to produce emergent outcomes that are difficult (or impossible) to predict from any individual element alone or the individual interactions (Miller & Page, 2007). The *complex* part of CAS refers to the interconnectedness of

H. E. Campbell et al., *Green Gentrification and Environmental Injustice*,
https://doi.org/10.1007/978-3-031-65100-7_3

these systems and the *adaptive* part refers to the ability or behavior of elements or agents that can change depending upon their environments (Desai, 2012). CAS can exhibit deep stability or unexpected surprises. These behaviors of CAS are not imposed from outside (except for things like meteorites hitting the earth or natural disasters destroying cities), but rather are the result of the internal self-organization of the myriad elements that make up the system.

CAS are also a system of systems, which refers to a collection of multiple distinct systems, integrated into a larger system with unique qualities. Cities dramatically exhibit this. As modeled by Campbell and Corley (2012) in Fig. 3.1, a

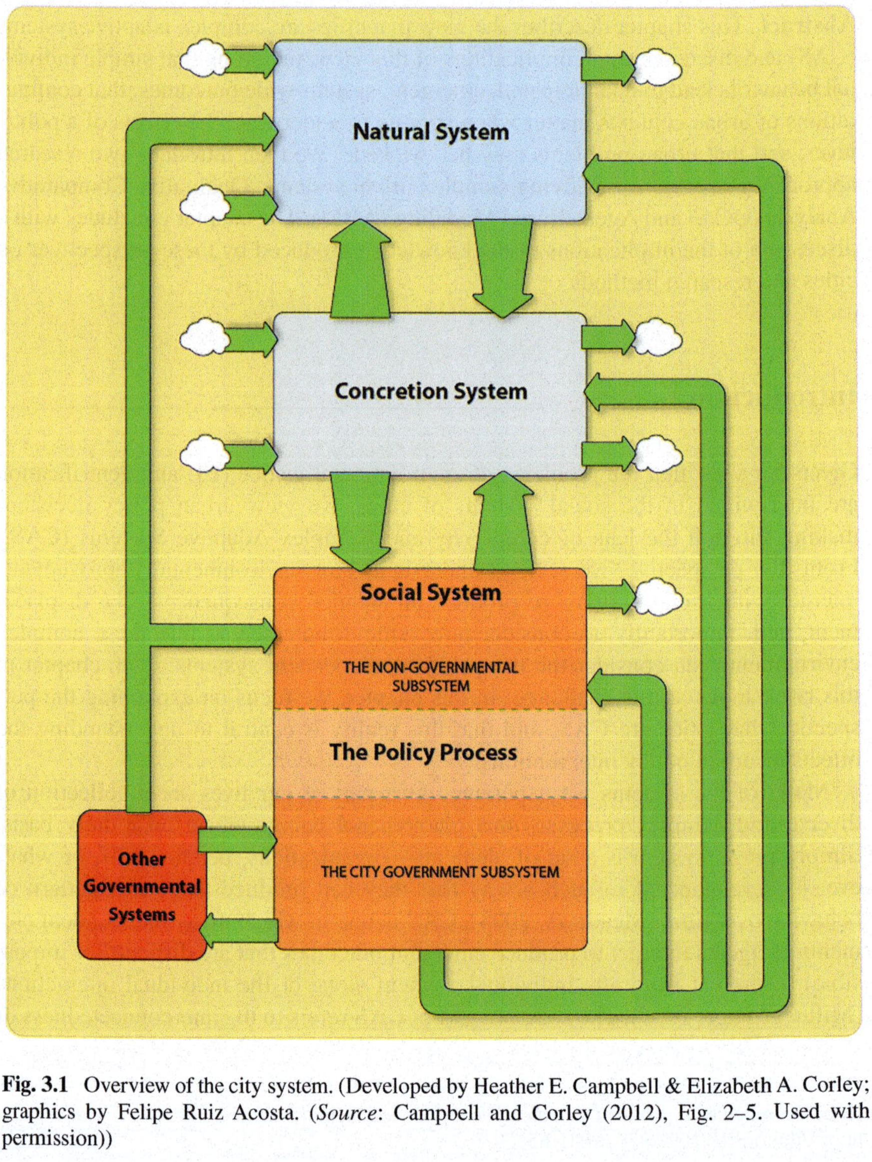

Fig. 3.1 Overview of the city system. (Developed by Heather E. Campbell & Elizabeth A. Corley; graphics by Felipe Ruiz Acosta. (*Source*: Campbell and Corley (2012), Fig. 2–5. Used with permission))

city includes Concretions (e.g., buildings, road networks), which are themselves systems; and Social Systems (e.g., markets and governments); all embedded in the Natural (environmental) System—which itself incorporates other systems such as water and weather. Any city is a multiplex CAS, consisting of a whole city as a CAS and the individuals, organizations, and environments that make up the city, each of which is also a CAS (Portugali, 2016). In such cities, urban change can take place through various factors, such as randomness, historical accident, physical determinism, natural advantage, comparative advantage (Batty, 2007), and intentional human interventions such as via policies. All contribute to the complexity we observe or experience from cities and neighborhoods.

CAS present *emergence*, which refers to the way in which a system undergoes a process of self-organization, transitioning from the properties of its components to a collective structure with unpredictable properties (Batty, 2007). Some commonly mentioned examples of emergence, in which the actions of agents result in system-level patterns, are flocks of birds, swarms of bees, standing ovations, traffic jams, and segregation (Resnick, 1994; Bedau & Humphreys, 2008; Schelling, 1978). In these examples, understanding the action of one agent (bird, bee, audience-member, driver, resident) cannot generally allow us to predict the emergent outcome; though formed by individual actions, in some cases the agents don't even desire the emergent outcome. The dynamic behaviors of agents and the structure of the system can create these unusual macroscopic structures.

In cities, researchers consider segregation (Schelling, 1978) and urban agglomeration (Wang et al., 2020) as emergent phenomena. We have argued that environmental injustice itself—a race-based environmental quality gap—is an emergent phenomenon (Campbell et al., 2015). The emergent property of CAS can arise from agents who rely on local information in the transitioning process without knowing its consequences to the system. Agents' adaptive (or learning) behavior can cause a change to the flow or structure of the system or create new adaptive behaviors of agents that take advantage of the features of the system (Nel et al., 2018). The decision of agents at a given point in time can lead to unexpected systemic behaviors because there are complex and dynamic relationships within the system, and also systemic inertia—as noted by Meadows (2008), up to a point, systems tend to return to their pre-shock state—that make it difficult for the decisions of agents to be immediately reflected in the behavior of the system. The agents within a CAS rarely make decisions at the same time within the same decision-making conditions; effects may not be proportional to causes due to latent processes and information loss in the processes; and there may be a time lag between the cause and the effect due to cascading processes (Kim & Maroulis, 2018). Effects are not necessarily driven by top-down plans and policies, but by the unfolding processes based on the interactions and interdependencies between autonomous actors that make up the system. This process can surprise us with the emergent behavior of the system, which can seem far from observable individual behavior.

CAS is a dynamic system of many interdependent and interacting components (agents, materials, subsystems) as a whole. In such a system, for a single actor (even government) to control or even coordinate the action of others as it intended may not be possible. This does not mean despair for science and policy research; it opens the search for new perspectives and approaches. Below we elaborate this perspective using the example of neighborhood change in cities and how CAS researchers approach such a system.

Complex Adaptive Systems View

Neighborhood Change

Urban systems have long been noted for their complex adaptive nature (Gallotti et al., 2021; Sengupta et al., 2016; Dong, 2023). Notably relevant to our approach here is work investigating the always-changing nature of urban housing markets and communities (Glaeser et al., 2023). In the US, the housing market is characterized by a relatively constant churn of both physical land uses and the social characteristics of structures. This churn of housing has been observed both at the individual level, with people and businesses moving/opening/closing, as well as at the community level, in the sense that a neighborhood, which is more than the sum of its human and structural elements, has a certain (but changeable) character.

At the individual level, people have been observed to relocate fairly frequently, particularly people who rent their homes (See Chap. 1). The result of independent individual decisions about where to live can be racial/ethnic change at the community level. For example, Wilson and Taub (2006) compare several neighborhoods in Chicago, demonstrating how some quickly change from being populated predominantly by one ethnic group to another, while others remain ethnically stable for extended periods of time. Communities may experience long periods of stability, or they may go through relatively constant change, with people moving out and others moving in, which could induce demographic changes for the community. What makes the difference? How could we know?

In the US, we tend to find that change happens less frequently or at a slower pace in rural areas (Foulkes & Newbold, 2008), although rural areas have seen significant population loss in recent decades, suggesting that rural areas are not stable so much as hollowing out. Stability can be lengthy in suburban areas, although it is worth noting that suburban areas are not immune to population shifts, and scholars are now noting demographic changes in inner-ring (older) suburbs (Hanlon, 2008). We tend to see more rapid changes to neighborhood character in urban core areas, particularly in dense cities that are constrained geographically, but, even in those cases, there are examples of urban core areas that remain stable for decades (Rohe & Stewart, 1996; Schieman, 2009). In other words, residential churn, i.e., people moving in and out, is rather normal, but it is not often predictable exactly how or when that churn will lead to a demographic transition. All neighborhoods experience changes over time, but there is no particular reason to expect that we can know in advance what that change will be, or when changes—even policy-intended changes—might occur. Some neighborhoods may change little over time, despite substantial changes around them, while others might change quickly despite existing in a relatively stable environment (Brueckner & Rosenthal, 2009). As we know historically, even sustained and serious efforts to cause change are not always successful (Peters & Fisher, 2004).

Neighborhoods are never completely unchanging and static, although some have shown to be stable for substantial lengths of time. Either can be beneficial or problematic; stability can bring comfort and a level of certainty to residents, but change can infuse communities with new amenities and increases in property values. The concern over gentrification and displacement is that a disruption in underlying stability may be beneficial to the group of residents moving in, but harmful to the residents who were living in the neighborhood before the gentrification began, particularly if the change happens quickly (Freeman & Barconi, 2004). With urban policy, we are not necessarily seeking to stop communities from changing, but rather to ensure that the changes that do happen are consistent with what local residents want and do not leave them in the lurch via a neighborhood that changes rapidly and displaces them with little warning. In other words, we might seek a level of stability or a slower pace of change—recognizing that changes will occur, but ideally these changes would occur gradually with minimal disruption to existing residents.

In part, our rationale for treating cities as CAS is that in systems approaches we generally seek stability in systems that are functioning well, while recognizing that they are constantly going through an adaptive process; but we seek to change systems that are functioning poorly. If we take a cue from an ecosystem approach, species who are well adapted to their environment thrive as long as that environment is relatively stable. Although it will not be static—for example, each specific plant may change and die—as long as the changes that occur are not disruptive, then the species will adapt with the environment. Disruptive changes—for example, if all the plants of a particular type die—like external shocks, put species at risk because they cannot adapt quickly enough.

But note the flipside of this analogy. If an ecosystem is degraded, keeping it stable is not desirable. Neighborhoods embedded with unhealthy levels of pollution are degraded environments for humans and put the local humans at risk. Therefore, city planners and others have considered how quality-of-life improvements can be made, while mitigating the potential for displacement. Neighborhood improvement, especially via amenities, safety, and accessibility, are facilitated by government, particularly via state and local investment. That is, we seek to use policy to keep healthy communities stable, while trying to break the stability of communities that are not doing well.

The facts that people move and neighborhoods change—natural processes in city as CAS—nonetheless present challenges for policy and planning. While gentrification and displacement are often seen as problems in cities, they are also the results of processes that a city cannot stop. If people have freedom of choice to relocate, then there is risk of gentrification[1]—and, indeed, evidence from other countries suggests that even making it illegal will not stop such change. We do not want to stop people from moving or neighborhoods from changing–but we do want to try to reduce the costs of such changes, especially for those who have already been the recipients of spatial, systemic social injustice via pollution.

Owing to the prior several decades of disinvestment, many urban policies of the 1990s were explicitly aimed at bringing the middle class back to the urban core (Hyra, 2017). Some revitalization projects funded by governments demolished public housing in neighborhoods and provided opportunities for the development of new market-rate housing (Wyly & Hammel, 1999; Goetz, 2003). From one perspective, these policies have been wildly effective; via the infusion of more amenities that increase the quality of life in the city, urban core neighborhoods began to change, and often change rapidly, with escalating property values. However, these benefits tended to accrue unevenly, with the biggest benefits going to the wealthy and the biggest costs to the poor, particularly the minority poor. All these contribute to realities we experience in cities.

CAS Research Framework

Above, we have briefly described neighborhood change dynamics in cities from the CAS perspective. Here we further discuss how researchers frame cities or urban systems as CAS for research. Initially, the researcher must delineate the boundary of a system of interest (Meadows, 2008; Richardson, 2011). An urban system as CAS can be conceptualized in a very generic and abstract way, as we describe neighborhood change above, Schelling's early work (1978), or our previous work on environmental justice (Campbell et al., 2015). Alternatively, an urban system can also refer to a particular city, like Chicago, which comprises the material and human components of Chicago while disregarding those of other places such as Indianapolis in Indiana or Milwaukee in Wisconsin, or it could model the entire region encompassing these cities. Modeling can specifically focus on just the energy systems of a specific city (e.g., Chicago), or the interlinked energy systems of multiple cities. In this case, the system delimits the significant aspects of energy production and consumption within a specified geographic area and omits less significant elements, such as the local art-education system. Depending on the question at hand, the energy system and the art-education system may be relatively separable subsystems of an urban system defined with a geographic boundary. When their coupling of subsystems is loose enough, interrelated systems can be conceptually broken down into smaller, independent parts without compromising the overall integrity of the large system (Ostrom, 2007). Defining the boundary of a system also includes

[1] And also a risk of depopulation and blight, of course.

defining temporal (e.g., 2010s) and spatial (e.g., highways, density) boundaries (Nel et al., 2018). For modeling purposes, these decisions on which boundaries of a system to consider and delineate depend upon a research question. It is also crucial to recognize that a CAS under consideration for research is an open entity that can, in reality, interact with elements outside of our designated boundary.[2]

Once the temporal, spatial, and topical boundaries of the system are delineated, it is necessary to identify which of the many elements that make up the city are appropriate for examination. We can draw on existing research or empirical knowledge to identify the basic building blocks, behaviors, characteristics, and structures of the urban system. The agents that we include in a model are driven by the goals of our inquiry. In our previous study of environmental justice, key agents included majority or minority residents, pollution-producing and non-pollution-producing firms, and sometimes local government within a growing artificial city (Eckerd et al., 2012). The interactions and interdependencies between agents can be represented through processes or structures (or networks) through which information or resources are exchanged or decisions are negotiated. For example, we modeled the processes of residential choices, firms' siting decisions, and government zoning scenarios in the aforementioned study.

For urban planners and policymakers working on a range of issues in cities, the ability to view urban systems through the CAS lens has implications. First, as noted above, problems that appear in CAS can be understood as emergent qualities generated by the dynamics of the system's components and the interdependence of different agents' interests or behaviors, rather than as a fixed system structure or permanence (Dong, 2023). For example, gentrification-induced displacement can be thought of as a phenomenon that can manifest differently in different contexts and conditions, rather than as a fixed structure that inevitably appears in all urban systems or that remains semi-permanent once it emerges. Second, sometimes such phenotypic features can be changed by making small changes to the processes and structures that drive interactions and interdependencies at the micro level. For example, highway traffic may be mitigated if all drivers are willing to allow enough space between their car and the ones around them, or even by changing the shapes of curves, or the narrowness of roads, or changing access to the highway. Environmental injustice or gentrification-induced displacement is a different type of issue than traffic flows and traffic jams, but it is worthwhile to try to improve our understanding of these important social matters using the CAS lens.

The orientation that one uses to craft policy approaches for a CAS may also differ from a traditional policy orientation. A CAS approach to policies is more likely to focus on ways to find balances in systems rather than trying to solve (what we recognize are intractable) problems. Using policy, we might influence what happens in a CAS, but we are unlikely to be able to control or thus solve the problem. In a sense, then, efforts to prevent gentrification are prone to failure because prevention assumes control and gentrification cannot be completely controlled. The

[2] The cloud graphics in Fig. 3.1 indicate this reality.

community will change at some point, and it is likely that nothing can be done to prevent that ultimate reality. Nor should we want to; cities should not be converted to flies in amber, even if we could—part of what makes cities valuable is their dynamics. Although housing churn has downsides, it also is an important engine of quality improvement (and sometimes wealth-building for the non-wealthy), and, while seeking to prevent housing churn may limit the effects of displacement, it may also thwart the environmental improvement of cities.

Recognizing that cities are CAS leads us to confront that our policy efforts to affect community stability are subject to sets of other circumstances that may or may not lead the policy changes to function as expected. This is not to suggest that we should not instigate policy efforts, of course, but rather to note that social efforts to suppress gentrification are likely to be as variably successful as history has shown our efforts to spur economic development to be. From this perspective, then, any preventative measures are subject to significant uncertainty. We might be instigating environmental improvements that are not realized by the residents who currently live in a community. Perhaps even worse, we might be failing to provide policy benefits to communities that need them for fear of a gentrification risk that may not materialize. Given the clustering of minority residents with pollution, declining to improve polluted neighborhoods due to fear of displacement means that we significantly limit our ability to clean our cities. Ultimately, we do not know when an improvement will spur gentrification, and when gentrification will lead to displacement, and when it will not, because whether it occurs or not is not independently caused by the policy change. Rather, there are myriad other factors interacting in the urban system to create sets of circumstances that may point in one direction or the other.

Urban researchers who seriously acknowledge the complexities of cities describe the problem of planning and managing cities as being consistent with the nature of wicked problems discussed by Rittel and Weber (1973) and other scholars in the 1960s and 1970s (Sengupta et al., 2016). "Wicked problems" are problems with no definitive solutions (Churchman, 1967; Rittel & Webber, 1973). Policy problems are wicked when the problem is difficult or impossible to solve with optimal solutions, or when implemented solutions create another set of problems, which might be more difficult to solve than the one addressed in the first place. Further, there is often no definitive way to describe these problems. The problem can look very different depending upon the perspective that one uses to view it, or the boundaries one chooses (as discussed above), and understandings of key concepts are often subjective. For example, the problem of climate change might look very different if viewed from a global perspective versus a local coastal region perspective, and also if viewed from an economic rather than an environmental lens. Definitions of abstract, but crucial, concepts like equality, justice, or fairness are subjective and unlikely to be agreed upon. We argue that the complexity of economic development, cleaning, greening, gentrification, and displacement present us with a wicked problem. The ramifications of the wicked policy problem and the attempt to solve that problem may reverberate across multiple policy systems with unintended and unanticipated effects.

The next question, then, is what methods can help us understand key elements of the urban CAS around emergent issues of environmental injustice, green gentrification, and displacement?

Two Research Approaches for the Urban CAS

Cities are complex adaptive systems (Batty, 2007), and therefore replete with wicked problems. Churchman (1967) described wicked problems as a "[…] class of social system problems which are ill-formulated, where the information is confusing, where there are many clients and decision-makers with conflicting values, and where the ramifications in the whole system are thoroughly confusing (p. 141)." In 1973, Rittel and Webber stated, "the search for scientific bases for confronting problems of social policy is bound to fail, because of the nature of these [wicked] problems" (p.155). Wicked problems abandon linear causality, and it is non-linear causality that is appropriate for unravelling wicked problems (Domegan et al., 2017).

Many methods we have used and still use to approach these problems—even standard logic itself—are inherently linear. Wicked problems cannot be solved with standard policy-analytic approaches; the interacting systems and nonlinear dynamics of complex systems mean that solutions based on such approaches often lead to unintended consequences. Fortunately, since Rittel and Webber were writing, the social sciences have begun to embrace and explore analysis techniques that were not available in the 1960s and '70s. Many of these analytical techniques and tools were developed to capture, understand, and explore the connectedness of parts and the surprising behavior of the whole system because many wicked problems are in fact emergent problems.

In this situation, what we need is not an approach that allows us to study all social problems in a specific way, but a framework that allows multiple tools and techniques to help us understand them. For instance, we can take the CAS perspective on cities and expand our knowledge by relying on different research approaches that embrace this perspective. While different approaches can be used to implement the CAS perspective, the key to them is to capture "interconnectedness" (i.e., underlying mechanisms of CAS) and understand complexities better. In this book, we use two research approaches to understand the complexity of urban systems: Qualitative Comparative Analysis (QCA) and Agent-based modeling (ABM). QCA has been developed to capture the complexity of social realities within the tradition of qualitative case studies. This approach focuses on exploring the unique sets of conditions (i.e., configurations) that produce the outcome of interest (Schneider & Wagemann, 2012). ABM has also been used as a tool for analyzing within the CAS perspective. This approach focuses on modeling agent-level (or micro-level) processes and helps us examine under what conditions a system's macro phenomena arise or do not.

Our approach recognizes the interacting systems that make up these wicked problems and identifies how sets of conditions—not individual factors one at a time—combine to generate the emergence of environmental injustice in cities. In short, these approaches are not aiming to estimate the average effect of an independent variable, assuming everything else holds constant and fixing time and space. The advantage of these CAS-responsive approaches from a policy point of view is that they provide new ways of structuring the complexity of the system (or policy problems) for better understanding of the system.

Qualitative Comparative Analysis (QCA)

QCA is a developing research approach to explore *configurations*. In the field of organization research, the study of configurations had a promising start in the 1970s and 1980s but had stalled until QCA brought the rebirth of academic interest (Miller, 2018). Miller articulates configurations as "common alignments of elements" (Miller, 1996, p. 506) and "commonly occurring constellations of often complementary elements" (Miller, 2018, p. 455). Miller (2018) also states that early organization research on configurations in the 1970s was a reaction to contingency theory and a desire to develop the mid-range approach linking the theory to empirics, and therefore focused on conceptual typologies and cluster analyses. Recent QCA research examines the co-presence of variables or attributes to theorize the rich complexity of relationships that characterize social phenomena, and also focuses on contrasts across configurations to explain specific outcomes (Greckhamer et al., 2018).

QCA's causal paths (i.e., configurations) capture the hidden interactions between conditions that produce or prevent the emergence of a social phenomenon. Causal paths are specific local mechanisms that are bounded by space and time, producing a specific social reality (Verweij, 2015). Schneider and Wagemann (2012) state that QCA is not just another data analysis technique, but it is a research approach. Thomann and Maggetti (2020) also suggest distinguishing QCA as a data analysis technique from QCA as a research approach because QCA encompasses various research-design issues that appear in processes before and after the analysis of the data. QCA includes three distinct features (Thomann & Maggetti, 2020; Verweij & Trell, 2019):

1. QCA is a mode of reasoning, which aims to help researchers generate new insights for a further theoretical development or a reexamination of existing theories through a configurational approach.
2. QCA examines cases, as does the traditional case-oriented approach, but also incorporates a condition-oriented approach in that it focuses on comparing cases using a set of conditions. Relying on cases and conditions together, QCA identifies combinations of conditions.
3. QCA identifies the local causal mechanisms in the form of configurations that explain how a certain outcome was produced or deterred rather than quantitatively estimating the effects of independent causes.

As for any research approach, certain types of questions are appropriate for QCA, but other types are not. QCA is most suitable to explore causal complexity underneath social realities, which refers to conjunctional causes, equifinality, and causal asymmetry (Schneider & Wagemann, 2012). To explain these three terms, let's assume that there are conditions (A, B, C, D, F) that have been considered and discussed to explain an outcome condition (Y). Our empirical analysis using QCA as a technique may produce the following hypothetical solution: (A*B) + (B*C) + (D* ~ F) → Y. This says that (A **AND** B) **OR** (B **AND** C) **OR** (D **AND NOT** F) are sufficient to produce the outcome, Y. In QCA, "*" indicates "Logical AND," "+" indicates "Logical OR," and "~" indicates "NOT."

The hypothetical solution indicates that there are three explanatory paths (A*B, B*C, D* ~ F). First, when there are multiple paths that can equally and sufficiently produce an outcome, which is noted using Local OR(+), this is called *equifinality*. Second, within each path, more than one condition is combined using Logical AND (*): i.e., *conjunctional causes* are allowed. Third, QCA assumes asymmetry of concepts and relations: i.e., QCA allows *causal asymmetry*.[3] This last assumption requires two separate definitions and operationalizations of a concept and relations in QCA. For example, in QCA, richness is not the opposite of poverty, as it would be viewed in Economic measures where "richness" and "poverty" are simply two parts of the same measure (e.g., income), one lower and one higher, and therefore the same causal variable causes increases or drops in the variable. In QCA these concepts need to be separately defined and treated as two different concepts because richness and poverty are two qualitatively different economic states. Additionally, sets of explanatory conditions identified for being rich as an outcome can be different from sets of explanatory conditions identified for not being rich as an outcome. Thereby, QCA aims to find the combinations of causal conditions (sets) that are necessary or sufficient to bring about the social phenomenon—or to prevent it—by systematically comparing cases and examining set relations of conditions (Schneider & Wagemann, 2012).

The characteristics of causal complexity are interlinked and integrated into the notion of necessity and sufficiency. If only a single condition is present for the outcome condition to occur, that condition is necessary and sufficient to cause the outcome condition by itself. Most of the time, social reality does not occur due to a single condition, but instead within the context of other conditions. For example, Kim and Verweij (2016) suggest that no single condition was necessary and/or sufficient for US states to adopt environmental justice policy. They found that the high racial diversity condition of a state is contained in all combinations linked with the policy adoption outcome, but that it cannot "produce" the outcome alone. In other words, the high racial diversity condition is insufficient [I] on its own for the outcome, but it is necessary [N] to form a sufficient configuration together with other conditions (i.e., a combination of democratic control of a state and weak interest group, or a combination of strong interest group and lower pollution level). In turn, neither of the two paths is necessary [U], but both are sufficient [S] to explain the adoption of a high level of environmental justice policy by states in the US. Such

[3] Causal asymmetry means that if (A*B) → Y, that does *not* imply that (~A* ~ B) → ~Y

INUS conditions—where some elements are Insufficient [I], but Necessary [N], but no one configuration is necessary [U], but more than one is sufficient [S]—are frequently observed in empirical studies using QCA. Causal paths that QCA assesses are subtle and tailored for contexts (Ragin, 1987).

While QCA is a brainchild of an American sociologist, Charles C. Ragin (1987), its use has been growing in various academic fields, including spatial planning and related research areas. Verweij and Trell's (2019) systematic literature review of QCA in spatial planning research shows that QCA has been embraced primarily due to its sensitivity to context, its ability to use small- or medium-N cases, and, of course, its attention to causal complexity. For instance, Gerrits et al. (2002) examined the ways in which cities transform through urban development or redevelopment projects, urban renewal projects, etc., and the planning and policy recipes that drive those changes in German, Austrian, and Swiss cities. Their findings show that successful transformations can be explained by a combination of a high strategic character of the project with a leading role of public actors, or a combination of a high strategic character with brownfield redevelopment.

Some researchers use QCA from the perspective of more traditional case-based, in-depth qualitative research with small or medium numbers of cases, while others use QCA for semi-quantitative research with large numbers of cases. Regardless of how QCA is used, it is a useful approach for developing theoretical perspectives that allow us to develop or identify local causal mechanisms without losing sight of the complexity of the real world. Again, the end-product of using QCA is not an estimate of average effect of a causal factor on an outcome. QCA is specifically designed to reveal or understand clusters of variables (i.e., configurations).

Agent-Based Modeling (ABM)

Agent-Based Modeling (ABM) is a computational approach, in which, nowadays, an artificial world is created in the computer. Schelling's checkerboard-based model leading to the emergent outcome of segregation is an example demonstrating that an ABM can provide unique insight into social-science topics. Schelling (1978) proposed a simple micro-level theoretical concept about individual behavior—people like to live near at least some other people with similar personal characteristics—and its implementation generated apparently complex behavior at a macro level (Schelling, 1978; Batty, 2007). Each iteration of the simulation consists of families scanning their neighborhood and deciding to move or stay depending upon whether the number of similar neighbors is below or above their threshold. Schelling started the simulation with an initial random distribution and required a relatively small number of iterations to arrive at segregated neighborhoods. This model provides a powerful illustration of how one can obtain apparently complicated patterns that attain a robust equilibrium from the repeated application of simple rules (e.g., scanning neighbors and making residential decisions based on simple preferences).

The strength of ABM as a simulation tool is that it can embrace diversity and complexity in populations, policies, and outcomes (Desai, 2012). ABM consist of agents and rules. To develop an ABM, one must define the relevant agents; define constraints of the "world" that the agents inhabit; and define the rules for interaction among these agents during a simulation. Different types of rules can be set up in ABM. First, action rules model how the agents interact and are codified through computer algorithms. In our environmental justice ABM (Campbell et al., 2015; Eckerd et al., 2012; Kim et al., 2014), we specified the processes of residential choices and firm-siting decisions, which leads to interaction between residents and firms in a confined geographical space. Also, rules for rule-change can be introduced. These rules define how the rules change based on interactions; these elements of an ABM directly respond to Casti's (1996) insight that "adaptive agents are ready to change their rules in accordance with new information that comes their way, continually adjusting to their environment" (p. 769). In another study, we experimented with the effects of different zoning scenarios (proactive versus reactive zoning strategies) on the environmental quality gap between majority and minority residents (Campbell et al., 2014). The policy scenarios, especially the reactive zoning strategy, are tested as a way of altering the existing problem of environmental injustice.

This control over the rules and how they adaptively change allows the analyst to explore, albeit in a limited fashion, the potential effect of rule (policy) changes in a CAS. Flexibility in designing new rules in a simulation allows researchers to test their assumptions about the mechanisms underlying complex social phenomena and how they percolate through the system (Resnick, 1994), and to explore the consequences of interventions in CAS (e.g., Campbell et al., 2014). Researchers can modify rules and parameters to explore other assumptions in the simulated reality (Marcy & Willer, 2002) and gain insights for action.

Another important aspect of ABM is that time matters; this is quite different from other formal modeling such as the Economic model of comparative statics, which focuses only on the equilibrium points. ABM allows analysis of the *path* to equilibrium (if any). In ABM, therefore, analysis can yield insights for both theory building and practice through action rules as well as suggestions for how these rules might change through an adaptive process. ABM goes beyond the deductive analysis of closed systems to provide interactive analytic support for inductive reasoning about open systems (Bankes, 2002).

Repetitive and competitive interactions between agents, which allow agent learning, are features of ABM. Even a simple agent-based model can exhibit complex behavior patterns and provide valuable insights into the dynamics of the real-world system that it emulates (Epstein et al., 1996). Agents may be capable of evolving and allowing unanticipated systemic behaviors to emerge. Sophisticated ABM sometimes incorporates neural networks, evolutionary algorithms, or other techniques to allow realistic learning and adaptation (Bonabeau, 2002). When some underlying relationships are known or hypothesized, it is also possible to embed them in the model in the form of statistical or optimization models. ABM is a package of techniques to build an artificial system and analyze the system's behaviors.

Because the ABM must work within the computer, it is also a formal method, simply instantiated in computer code rather than in mathematical equations.

Since the 1990s, ABM has been widely applied in urban and geospatial studies (Chen, 2012). Its strong appeal in such studies comes from the ontological correspondence between computer agents in the model and real-world actors. It is easy to represent actors (e.g., commuters), physical or spatial environments (e.g., transportation), and their interactions in ABM tools. The intuitive descriptions of complex social systems and the technical tools to implement them has enabled a new understanding of social phenomena.

Due to the explosion of use of ABM in urban and related areas, several review articles specific to sub-topics are already published (see Chen, 2012, on urban studies; Parker et al., 2003, on land use and land cover; O'Sullivan, 2008, on geosimulation; Hare & Deadman, 2004, on environmental and ecological modeling). We don't review those here, but it is worth noting that ABM has gained strong currency as a research tool for cities and urban issues.

Simulations are not true experiments, so what we can deduce from a simulation is not always obvious. For instance, if the outcomes of simulations are based on a priori assumptions, then distinguishing between results that tautologically follow from the assumptions and those that emerge from the complex interactions of the assumptions and the rules that govern those interactions may not be straightforward. Also, there is no guidance on determining the appropriate granularity of an ABM to represent a system, and it is important to recognize that the researcher determines the appropriate level based on the purpose of the study. A formal Economic comparative statics model that analyzes only a few of the potential variables and only a few potential changes does not address the methodological issues of which variables were chosen and, in the same way, ABM does not provide answers to these methodological issues. However, ABM does broaden the spectrum of policy analysis and synthesis tools by bringing unique strengths to the process of modeling autonomous adaptive human behavior, interactions, timing, and connections to systemic patterns that have not been available via traditional analytical tools.

Conclusion

Many decades ago, Weaver (1948) indicated that the type of problems upon which academic areas such as biology, medical science, psychology, economics, and political science focus is *the problem of organized complexity*—that is, "problems which involve dealing simultaneously with a sizable number of factors which are interrelated into an organic whole" (Weaver, 1948, p. 539). Due to the nature of such problems, Weaver (1948) envisioned that these areas may not be successfully understood when only a small number of variables are considered or are explored by statistical techniques. Similarly, we have long known that public policy problems or urban planning problems are wicked or complex (Rittel & Webber, 1973). However, to make it manageable for policymakers or practitioners, we have often reduced them using a few variables or approximated them using statistical models that seek to eliminate dynamism and focus on all-else-equal results. Some scholars deal with these issues by turning to case studies, which retain the context surrounding causes and effects, but make it difficult to generalize. As a result, there is a mismatch between the way in which we (especially qualitative researchers) characterize urban policy issues and the tools we use (especially statistical researchers) to study them.

Recent advances in our understanding of complexity have given us a conceptual framework, language, and tools that provide us the capacity to approach such problems differently than through our habitual mindset and tools. While these advances do not provide tangible empirical numbers, as do deterministic models or statistics, they can help urban researchers systematically understand complex realities, acknowledge uncertainties, and think dynamically. For this reason, the research approaches used in this book are well-suited tools for understanding complex phenomena and, ultimately, for developing theory, and letting us move beyond local-knowledge intuition to configural and emergent understandings of such phenomena.

In this chapter, we considered what kinds of questions modeling the city as a CAS generates, frameworks that complexity-informed researchers have developed to study cities' complexity, and two research approaches that can be used to construct or leverage the connectedness that drives complexity for insights on wicked problems. In the rest of this book, we explore various ways that cities may use policy to affect the pace of change and reduce the risk that disruptive changes are likely to occur, rather than focusing on preventing changes from occurring.

References

Bankes, S. C. (2002). Tools and techniques for developing policies for complex and uncertain systems. *Proceedings of the National Academy of Sciences, 99*(3), 7263–7266. https://doi.org/10.1073/pnas.092081399

Batty, M. (2007). *Cities and complexity: Understanding cities with cellular automata, agent-based models, and fractals*. MIT Press.

Bedau, M. A., & Humphreys, P. (2008). *Emergence: Contemporary readings in philosophy and science*. MIT Press.

Bonabeau, E. (2002). Agent-based modeling: Methods and techniques for simulating human systems. *Proceedings of the National Academy of Sciences, 99*, 7280–7287. https://doi.org/10.1073/pnas.082080899

Brueckner, J., & Rosenthal, S. (2009). Gentrification and neighborhood housing cycles: Will America's future downtowns be rich? *Review of Economics and Statistics, 91*(4), 725–743.

Campbell, H. E., & Corley, E. A. (2012). *Urban environmental policy analysis*. ME Sharpe.

Campbell, H., Kim, Y., & Eckerd, A. (2014). Local zoning and environmental justice: An agent-based analysis. *Urban Affairs Review, 50*(4), 521–552. https://doi.org/10.1177/1078087413505736

Campbell, H. E., Kim, Y., & Eckerd, A. (2015). *Rethinking environmental justice in sustainable cities: Insights from agent-based modeling*. Routledge.

Casti, J. L. (1996). Lighter than air. *Nature, 382*, 769–770. https://doi.org/10.1038/382769a0

Chen, L. (2012). Agent-based modeling in urban and architectural research: A brief literature review. *Frontiers of Architectural Research, 1*(2), 166–177. https://doi.org/10.1016/j.foar.2012.03.003

Churchman, C. W. (1967). Guest editorial: Wicked problems. *Management Science, 14*(4), B141–B142. https://www.jstor.org/stable/2628678

Desai, A. (Ed.) (2012). *Simulation for policy inquiry*. Springer. https://doi.org/10.1007/978-1-4614-1665-4.

Domegan, C., HcHugh, P., Biroscak, B. J., Bryant, C., & Calis, T. (2017). Non-linear causal modeling in social marketing for wicked problems. *Journal of Social Marketing, 7*(3), 305–329.

Dong, M. W. Y. (2023). A critical analysis on complex urban systems and complex systems theory. *Journal of Computing and Natural Science, 3*(1), 24–34. https://doi.org/10.53759/181X/JCNS202303003

Eckerd, A., Campbell, H., & Kim, Y. (2012). Helping those like us or harming those unlike us: Illuminating social processes leading to environmental injustice. *Environment and Planning B, 39*(5), 945–964. https://doi.org/10.1068/b38001

Epstein, J. M., Axtell, R., & 2050 Project. (1996). *Growing artificial societies: Social science from the bottom up*. Brookings Institution Press.

Foulkes, M., & Newbold, K. B. (2008). Poverty catchments: Migration, residential mobility, and population turnover in impoverished rural Illinois communities. *Rural Sociology, 73*(3), 440–462. https://doi.org/10.1526/003601108785766525

Freeman, L., & Braconi, F. (2004). Gentrification and displacement New York City in the 1990s. *Journal of the American Planning Association, 70*(1), 39–52.

Gallotti, R., Sacco, P., & De Domenico, M. (2021). Complex urban systems: Challenges and integrated solutions for the sustainability and resilience of cities. *Complexity, 2021*, 1–15. https://doi.org/10.1155/2021/1782354

Gerrits, L., Pagliarin, S., Klin, K. U., & Knieling, F. (2002). Tracing complex urban transformations in Germany, Switzerland and Austria using trajectory-based qualitative comparative analysis (TJ-QCA). *Cities, 141*, 104507. https://doi.org/10.1016/j.cities.2023.104507. https://papers.ssrn.com/sol3/papers.cfm?abstract_id=4212361

Glaeser, E. L., Luca, M., & Moszkowski, E. (2023). Gentrification and retail churn: Theory and evidence. *Regional Science and Urban Economics, 100*, 103879. https://doi.org/10.1016/j.regsciurbeco.2023.103879

Goetz, E. G. (2003). *Clearing the way: Deconcentrating the poor in urban America*. The Urban Institute.

Greckhamer, T., Furnari, S., Fiss, P.C., & Aguilera, R. V. (2018). *Studying configurations with qualitative comparative analysis: Best practices in strategy and organization research.*

Hanlon, B. (2008). Fixing inner-ring suburbs in the US: A policy retrospective. *International Journal of Neighbourhood Renewal, 1*(3), 1–30.

Hare, M., & Deadman, P. (2004). Further towards a taxonomy of agent-based simulation models in environmental management. *Mathematics and Computers in Simulation, 64*, 25–40. https://doi.org/10.1016/S0378-4754(03)00118-6

Hyra, D. S. (2017). *Race, class, and politics in the Cappuccino City*. University of Chicago Press.

Kim, Y., & Maroulis, S. (2018). Rethinking social welfare fraud from a complex adaptive systems perspective. *Administration & Society, 50*(1), 78–100. https://doi.org/10.1177/0095399715587520

Kim, Y., & Verweij, S. (2016). Two effective causal paths that explain the adoption of US state environmental justice policies. *Policy Sciences, 49*(4), 505–523.

Kim, Y., Campbell, H., & Eckerd, A. (2014). Residential choice constraints and environmental justice. *Social Science Quarterly, 95*(1), 40–56. https://doi.org/10.1111/ssqu.12033

Marcy, M. W., & Willer, R. (2002). From factors to actors: Computational sociology and agent-based modeling. *Annual Review of Sociology, 28*, 143–166. https://doi.org/10.1146/annurev.soc.28.110601.141117

Meadows, D. H. (2008). *Thinking in systems: A primer*. Chelsea Green.

Miller, D. (1996). Configurations revisited. *Strategic Management Journal, 17*(7), 505–512.

Miller, D. (2018). Challenging trends in configuration research: Where are the configurations? *Strategic Organization, 16*(4), 453–469.
Miller, J. H., & Page, S. (2007). *Complex adaptive systems: An introduction to computational models of social life*. Princeton University Press.
Nel, D., Du Plessis, C., & Landman, K. (2018). Planning for dynamic cities: Introducing a framework to understand urban change from a complex adaptive systems approach. *International Planning Studies, 23*(3), 250–263. https://doi.org/10.1080/13563475.2018.1439370
Ostrom, E. (2007). A diagnostic approach for going beyond panaceas. *Proceedings of National Academy of Sciences, 104*(39), 15181–15187. https://doi.org/10.1073/pnas.0702288104
O'Sullivan, D. (2008). Geographic information science: Agent-based models. *Progress in Human Geography, 32*(4), 541–550. https://doi.org/10.1177/0309132507086879
Parker, D. C., Manson, S. M., Janssen, M. A., Hoffman, M. J., & Deadman, P. (2003). Multi-agent systems for the simulation of land-use and land-cover change: A review. *Annals of the Association of American Geographers, 93*, 316–340. https://doi.org/10.1111/1467-8306.9302004
Peters, A., & Fisher, P. (2004). The failures of economic development incentives. *Journal of the American Planning Association, 70*(1), 27–37.
Portugali, J. (2016). What makes cities complex? In J. Portugali & E. Stolk (Eds.), *Complexity, cognition, urban planning and design* (Springer proceedings in complexity). Springer. https://doi.org/10.1007/978-3-319-32653-5_1. http://www.spatialcomplexity.info/files/2013/10/Portugali.pdf. Accessed 20 Oct 2023
Ragin, C. C. (1987). *The comparative method: Moving beyond qualitative and quantitative strategies*. University of California Press.
Resnick, M. (1994). *Turtles, termites, and traffic jams: Explorations in massively parallel microworlds*. MIT Press.
Richardson, G. P. (2011). Reflection on the foundation of system dynamics. *System Dynamics Review, 27*(3), 219–243. https://doi.org/10.1002/sdr.462
Rittel, H. W. J., & Webber, M. M. (1973). Dilemmas in a general theory of planning. *Policy Sciences, 4*, 155–169. https://doi.org/10.1007/BF01405730
Rohe, W. M., & Stewart, L. S. (1996). Homeownership and neighborhood stability. *Housing Policy Debate, 7*(1), 37–81. https://doi.org/10.1080/10511482.1996.9521213
Schelling, T. C. (1978). *Micromotives and macrobehavior*. W.W. Norton.
Schieman, S. (2009). Residential stability, neighborhood racial composition, and the subjective assessment of neighborhood problems among older adults. *The Sociological Quarterly, 50*(4), 608–632. https://doi.org/10.1111/j.1533-8525.2009.01158.x
Schneider, C. Q., & Wagemann, C. (2012). *Set-theoretic methods for the social sciences: A guide to qualitative comparative analysis*. Cambridge University Press.
Sengupta, U., Rauws, W. S., & De Roo, G. (2016). Planning and complexity: Engaging with temporal dynamics, uncertainty and complex adaptive systems. *Environment and Planning B: Planning and Design, 43*(6), 970–974. https://doi.org/10.1177/0265813516675872
Thomann, E., & Maggetti, M. (2020). Designing research with qualitative comparative analysis (QCA): Approaches, challenges, and tools. *Sociological Methods & Research, 49*(2), 356–386. https://doi.org/10.1177/0049124117729700
Verweij, S. (2015). Once the shovel hits the ground: Evaluating the management of complex implementation processes of public-private partnership infrastructure projects with qualitative comparative analysis. *SSRN Electronic Journal*. https://doi.org/10.2139/ssrn.2646785
Verweij, S., & Trell, E. M. (2019). Qualitative comparative analysis (QCA) in spatial planning research and related disciplines: A systematic literature review of applications. *Journal of Planning Literature, 34*(3), 300–317. https://doi.org/10.1177/0885412219841490
Wang, C., Wang, Z., & Li, Q. (2020). Emergence of urban clustering among U.S. cities under environmental stressors. *Sustainable Cities and Society, 63*, 102481. https://doi.org/10.1016/j.scs.2020.102481

Weaver, W. (1948). Science and complexity. *American Scientist, 36*(4), 536–544. https://www.jstor.org/stable/27826254

Wilson, W. J., & Taub, R. P. (2006). There goes the neighborhood: Racial, ethnic, and class tensions in four Chicago neighborhoods and their meaning for America. New York: Knopf.

Wyly, E. K., & Hammel, D. J. (1999). Islands of decay in seas of renewal: Housing policy and the resurgence of gentrification. *Housing Policy Debate, 10*(4), 711–771. https://doi.org/10.1080/10511482.1999.9521348

Chapter 4
Conceptual Clarification: Gentrification, and Displacement

Abstract This chapter elucidates the concept of gentrification as a complex process and endeavors to explicate what makes the issue complex. An advantage of clarifying the concept is the ability to distinguish the triggers, stages, and outcomes of gentrification. More importantly, it facilitates the observation of the unintended consequence of the gentrification process, namely displacement, independently from the process itself. This chapter ends with insights gained from our endeavor to clarify this convoluted concept.

Introduction

As we discussed previously, the issue of gentrification was noticed in London, UK, decades ago—an influx of the "gentry" to a low-income neighborhood during the 1950s and 1960s (Zuk et al., 2018)—but how the issue has unfolded depends upon contexts (different locations and times). For example, we discussed that gentrification in the US is more common in recent decades (since 2000), and we also see different characteristics of the people who move in and are forced out in major cities (Hwang & Lin, 2016). Additionally, the process can be triggered not only by people moving in, but also by investments by governments and developers, who want to develop an area for a public or private purposes. The concept of gentrification has also evolved to encompass other variants, such as green gentrification or climate gentrification (Anguelovski et al., 2022; Shokry et al., 2020). In short, "gentrification scholars regularly refer to displacement, but equally apply a variety of overlapping and related concepts that they sometimes appear to regard as synonymous" (Elliott-Cooper et al., 2020, p. 493). The use of the term gentrification has become confusing, offering various interpretations or applications in different literatures.

This chapter clarifies key concepts of this book, gentrification and displacement, from the complexity perspective. Gentrification is a process that includes how the interaction between people and environment can lead to an undesirable consequence in a neighborhood—displacement (Zuk et al., 2018). To comprehensively outline

H. E. Campbell et al., *Green Gentrification and Environmental Injustice*,
https://doi.org/10.1007/978-3-031-65100-7_4

the process of gentrification while emphasizing its crucial aspects, two distinct gentrification processes, which are tightly intertwined with views about the initial driver of the process, are first presented (demand-side versus supply-side). This is followed by a breakdown of gentrification processes into its various stages (gentrifiable, gentrifying, and gentrified). Finally, we make a conceptual differentiation between gentrification and displacement: *gentrification* as a neighborhood change process associated with the improvement of built environments but also incubating the potential of existing residents' displacement, in contrast with *displacement* specifically as an unintended consequence of the gentrification process that may or may not be observed in a gentrifying or gentrified neighborhood. Our concern is certainly the occurrence of significant displacement (specifically of environmentally disadvantaged groups) induced by the gentrification process.

Gentrification as a Complex Process

Early gentrification literature addressing the US context was mostly about describing the new process of wealth returning to inner-core urban areas after decades of suburban development and urban disinvestment. The idea initially emerged from a critical perspective, as implied by the terminology that has since taken hold: *gentrification*. This initial definitional phase was characterized by a critique of the influence of capitalism in housing, with the process often described as an invasion of the wealthy into poor neighborhoods–sometimes including war metaphors (Hamnett, 1991; Eckerd & Reames, 2012).

As scholarship became more empirically based, two early frames for explaining the process emerged, one centered on supply and the other on demand factors. The demand focus, characterized by David Ley (1986), explored economic factors and incentives that drew the wealthy back to the urban core. Issues like traffic congestion, preferences for walkability or otherwise short commutes, and access to urban amenities drew the wealthy to urban core areas so that residents could live in closer proximity to jobs and leisure activities. The supply argument, most prominently proffered by Neil Smith (1987), focused on housing market considerations and the "rent gap" that opened after decades of urban neglect. Under this hypothesis, wealthy interests were drawn to urban areas because the housing stock retained latent value despite years of neglect; developers could recognize opportunities to purchase undervalued homes, rehabilitate them, and realize substantial profits after resale.

Much of the literature through the last part of the twentieth century flowed from one of these two premises. Much of the economically oriented literature focused on demand-side arguments, or blended the demand and supply arguments together, to create a picture of what the process looks like. There is a widespread understanding of the nature of what happens–gentrification occurs as demand for housing in a previously poor community increases. Often there is some initial investment that occurs that makes the neighborhood more desirable; this could be due to policy changes or market factors.[1] This desirability is reflected in a combination of more incoming residents, further investment in the form of housing and/or commercial renovations, which further increases desirability and demand. This cycle continues to reinforce itself until housing prices and rents begin to rise exponentially, pricing out the neighborhood's original residents (Hamnett, 1991; Freeman, 2005). Renters are often affected first as they can no longer afford to pay escalating rental prices, and homeowners can be affected subsequently through offers to purchase homes the sale of which can create generational wealth and/or through property taxes that become unaffordable due to updated home price assessments (Ellen & O'Regan, 2011).

Figure 4.1 includes the two different processes of gentrification as discussed above. Blue symbols in the figure illustrate the situation that gentrifiers (people) move into a previously undesirable, segregated, and low-income neighborhood and trigger the gentrification process (demand-oriented process). Orange symbols illustrate that gentrification may be triggered by significant investments to improve the built environment by governments or developers (supply-oriented process). Such investments can also trigger the gentrification process. The figure implies that some neighborhoods may experience gentrification through the path of demand-oriented factors, other neighborhoods may do so through the path of supply-oriented factors, and still others may experience both paths to gentrification simultaneously.

Regardless of triggers, the benefit of neighborhood improvement and investment comes with the cost of increases in property values (and thus property taxes) and rents. We do not want to lose sight of the fact that these "economic development" factors can have real benefits to all residents. However, the crux of the gentrification-related displacement problem is *significantly increased rents and property values in neighborhoods* which *may* create an affordable housing crisis. This is an opportunity for those who can afford the change (better-off existing residents and new

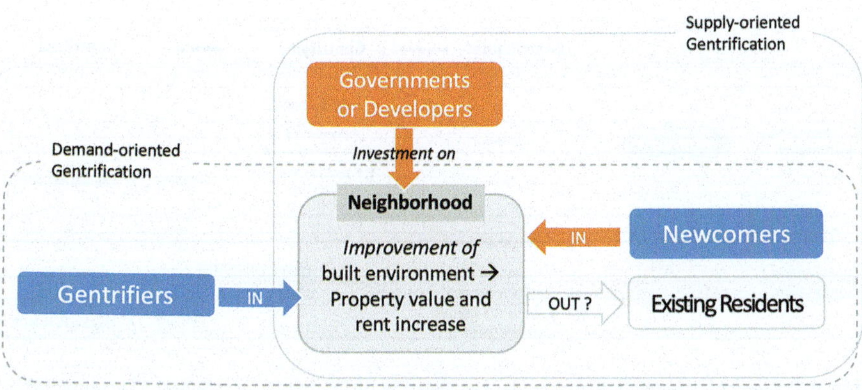

Fig. 4.1 Gentrification processes. (*Source*: The authors)

[1] This can also occur when a nearby more-desirable area becomes denser and more pricey

residents) to enjoy the benefits of improvement, but this also forces out those who cannot afford the change from their own preferred neighborhoods (worse-off existing residents). A gentrification-displacement cycle begins, where property values escalate, displacing even more people, until, in the end, the community becomes wealthy but with very little of that wealth being accrued by the original residents—that is, the wealth moves from outside the community into the community rather than the wealth being derived within the community. But, as we mention in Chap. 1, the gentrification process may not always result in significant resident displacement. Thus, the gray box labeled "Existing Residents" for outflow migration in Fig. 4.1 may be present with varying levels of displacement in a gentrified neighborhood. At one extreme, there may be an increase in the density of the area without noticeable displacement. At the other end, the process may result in the significant displacement of existing residents.

Gentrification Phases

The process can be conceptualized with at least three distinct phases: gentrifiable, gentrifying, and gentrified (see Fig. 4.2). A gentrifiable neighborhood usually is a neighborhood that is, at the very least, comparatively poorer than other neighborhoods in the region, and other considerations may also be relevant such as the educational levels or ethnic or racial makeup of the neighborhood before the gentrification process begins. That is, gentrification is generally not considered possible in neighborhoods that are already comparatively well off.[2]

Once gentrification begins in a gentrifiable neighborhood, the neighborhood undergoes the process of changes (gentrifying) until it slows down and reaches a plateau of higher wealth (gentrified). The quantitative studies of gentrification often

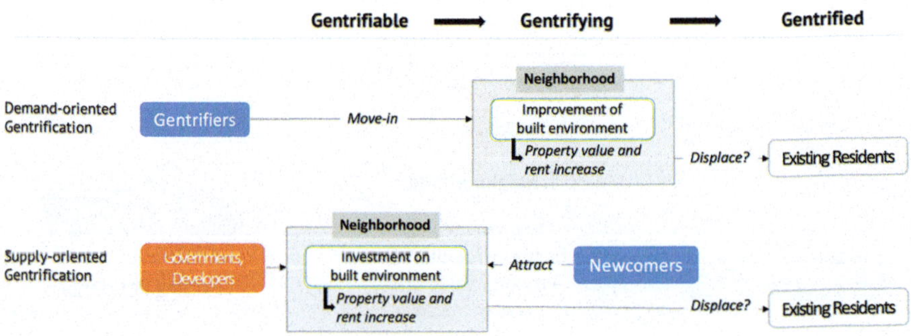

Fig. 4.2 Gentrification and displacement processes in different phases. (*Source:* The authors)

[2] In the New York City context, there is a case presented in which a wealthy neighborhood becomes even more wealthy, with the top 5% being displaced by the top 1% (Gould & Lewis, 2017). Because of our focus on EJ, we aren't really concerned with such situations.

take advantage of the phases to measure gentrification as an outcome. For example, empirical studies identify gentrifiable neighborhoods at the beginning of the study's time point and then identify those gentrified at the end time (e.g., Chapple et al., 2017; Stuhlmacher et al., 2022). In between, the process of gentrifying is assumed. This stages framework, or empirical studies relying on the framework, still do not address the issue of a time horizon of gentrification. It is not clear from our scan of the gentrification literature what the most appropriate temporal context is. It is clear that gentrification and displacement take some amount of time to fully process, but the time horizon in any particular neighborhood or region can be highly variable. For us, one concern is whether the gentrification-and-displacement timeline is long enough to allow existing residents to benefit from any cleaning and/or greening, or not.

Figure 4.2 shows very clearly that gentrification involves changes in both people and environment, but how they interact in the gentrification process may be different, depending upon phases. In the demand-oriented gentrification process, built environment improvements and housing price changes may be observed at the moment of a gentrifying stage after gentrifiers move into a gentrifiable neighborhood. Existing residents can feel displacement pressure and risk of moving, but the neighborhood's demographic changes may not be observed at the gentrifiable phase. Although the degree and speed of demographic changes may be different depending upon neighborhoods, we may actually observe displacement at the gentrifying stage. Such a demographic change may plateau once a neighborhood is gentrified, when the process has stabilized. In the supply-oriented gentrification process, governments or developers may invest in a neighborhood when it is gentrifiable for various purposes (e.g., adding amenities), which increase property values and rents in the neighborhood (although, of course, that may not be the intention of the policy). Unlike the demand-oriented process, at the gentrifying phase, newcomers who can afford increased living expenses may be attracted to the remediated neighborhood. Existing residents may be displaced when the process is completed and demographic change has plateaued.

Gentrification-Led Displacement

Our interest in gentrification-induced displacement stems from our longstanding commitment to environmental justice. Elliott-Copper et al. (2020) accurately noted that gentrification can lead to eviction, expulsion, and exclusion at various scales and speeds, ultimately severing the link between individuals and their environment. This severed connection between individuals and their environment can be most precisely conveyed through the term displacement rather than the term gentrification.

In the context of gentrification, the revitalization of urban areas through development can result in displacement, which refers to an involuntary cause for the displacement of incumbent residents; displacement as a series of events when a neighborhood gets too expensive for those who cannot afford the rising expense, especially with regards to housing costs. However, distinguishing between voluntary migration (choice-based outmigration) and involuntary reasons (especially related to the gentrification process) for displacement is not always straightforward (Lyons, 1996). Demographic changes can also occur through replacement (succession) rather than displacement, as noted by Hamnett (2003). We also acknowledge that displacement can occur via the direct removal of low-income households, but also via other indirect forms such as where longtime residents might not feel at home anymore in their changing neighborhood because of declining cultural identity (Marcuse, 1985). In reality, multiple reasons can drive residents' outflow migration simultaneously, and displacement can happen in much more subtle ways than some previous studies have operationalized and measured for gentrification. Still, the most clear form of involuntary displacement is when housing costs rise such that those living in the area before can no longer afford to remain. To us, this is especially problematic when those who have borne the costs of pollution are forced out before they are able to reap the benefits of cleaning and greening.

Our intention is not to ignore these complexities of neighborhood changes, but to explicate the complexities surrounding environmental improvement, gentrification and displacement into a framework that aids with conceptual clarification. While here we focus on the displacement of longtime residents due to gentrification in a neighborhood, we also acknowledge that not all urban displacements are associated with a gentrification process. Generally, it is challenging to determine the reasons for outflow migration in a neighborhood. Gentrification and displacement are clearly linked, and displacement is a defining feature of gentrification, but it is less clear how much of a defining feature it is, how and why displacement occurs, and over what timeframe it happens.

Much of the focus in the literature is on the drivers of the initial part of this process, and potential remedies or preventive measures that can ameliorate the latter part, displacement. To clarify the nature of the process, we describe what happens as two different (but obviously related) concepts are separated. Although it is relatively recent that these two concepts have been explicitly distinguished in gentrification research (Eckerd et al., 2019; Elliott-Cooper et al., 2020; Zuk et al., 2018), much of the literature has in fact been framed along these lines, with most of the causal empirical work focused on issues related to gentrification and much of the critical work focused on issues related to displacement (Freeman, 2005; Eckerd & Reames, 2012; Ellen & O'Regan, 2011).[3]

The gentrification oriented research has generally focused on the drivers of the process—notably, for our purposes here, policy-oriented antecedents, but also market-oriented drivers as well. There is good evidence to support both the demand and supply arguments about process initiation. In the demand case, gentrification has been associated with characteristics of the community, such as its proximity to

[3] One reason that the empirical work has focused on gentrification is that this is much easier to observe empirically than is the displacement piece. Americans, and especially poor Americans, move at a fairly high rate even absent gentrification.

the central business district (CBD), its transit accessibility, its proximity to environmental and recreational amenities, and the distinctiveness or architectural aesthetics of its housing stock (Eckerd et al., 2019). Demand for housing by wealthier residents has also been associated with changes that policy brings about, such as the construction of transit stations, the remediation of hazardous land uses, and the placement of desirable land uses such as parks or entertainment districts. Supply has also played a role, as the process has also been associated with specific efforts by developers to spur interest in a community, such as through establishment of arts-related spaces, restaurants, coffee shops, art galleries, and breweries, and the rehabilitation and renovation of housing done for the specific purpose of resale. In short, because urban settings are complex and interdependent, there are a wide-ranging number of factors that sometimes spark the gentrification process, but it has been difficult to pinpoint one or two specific things that make it happen. In some circumstances, almost any desirable change might spark gentrification, while in other circumstances all of these changes may not lead to the process. And, again, we wish to note that gentrification *alone*—absent displacement—may be desirable; in the US context with fairly efficient housing markets, any desirable addition to a neighborhood should cause some type of increase in housing values.[4]

Displacement-oriented literature tends to focus less on the characteristics of the neighborhood and more on the people that live there. Where the demand- or supply-oriented gentrification literature is frequently framed in economic or policy terms (Zuk et al., 2018), the displacement literature is more diverse and multidisciplinary, ranging from Marxist critiques of the housing market, social justice focuses on the racial component of the gentrification and displacement process, and mixed-methods research to assess what happens both to the nature of the community that is gentrified and to the residents that are displaced (Eckerd & Reames, 2012; Richardson et al., 2019). This scholarship typically focuses on some aspect of the injustice of displacement, not arguing that communities should never change, but rather arguing that current residents should have some say in what happens to their communities. This is a broad overview, however, as this research may come from geographers, sociologists, political scientists, or even anthropologists. The common theme is a focus on what is lost when gentrification happens and, to a lesser extent, on what can be done to mitigate the negative effects.

Interacting Processes

Gentrification as a process encompasses both changes in a neighborhood's environment and changes in the people who inhabit it. The recent scholarly examinations of gentrification have compelled researchers to disentangle the interlinked notions of

[4] We note that undesirable—from our perspective—changes, such as the decrease in minority residents, may also cause housing-price increases.

gentrification and displacement (Eckerd et al., 2019; Elliott-Cooper et al., 2020; Richardson et al., 2019; Zuk et al., 2018). For example, a few recent gentrification studies use machine-learning techniques and Google Street images to focus on understanding changes in the features of the built environment (e.g., changes in individual property, or spatial concentration of positive changes in the physical appearance of several properties in close proximity) (Illic et al., 2019; Naik et al., 2014, 2015). Even in these data-heavy studies of gentrification, researchers acknowledge that socioeconomic and cultural changes within and around a space are required in order to conform with what we think of as gentrification. The connection of gentrification and displacement is more complex than previously assumed within a single gentrification concept (Elliott-Cooper et al., 2020; Richardson et al., 2019). Conflating the two suggests that displacement is at the heart of the definition of gentrification rather than the crux being reevaluation of land and properties. This conflation has drawn our attention from an environmental justice perspective. We would like to see improvement of the neighborhood and its environment without the harms associated with displacement of those who have suffered a degraded environment. This leads us to the need to differentiate what are, after all, different components of a complex process.

Tracking people's movements and identifying reasons for resident outmigration pose significant challenges. Despite the potential for investments in built environments to increase the risk of displacement, we may observe varying degrees and rates (speeds across time) of displacement in gentrified neighborhoods during the same or different time periods. For example, if incoming newcomers are moving into newly built or vacant units, the process may result in densification without displacement. In other situations, however, displacement can occur so slowly that it remains unidentified until a couple of decades afterward. It becomes difficult to discern from the base-level population-churn in a community (or even from succession since some residents exit due to mortality). If so, in spite of change, incumbent low-income, or racial- and ethnic-minority residents may benefit from environmental improvements for a considerably long time span. Therefore, it is a valuable pursuit to distinguish displacement from the gentrification process, while acknowledging that the risk of the former can rise because of the latter.

Policy Implications

Efforts aimed at clarifying the intricate concepts of gentrification and displacement provide several policy implications and vantage points. Regardless of the disciplinary perspective, almost all gentrification literature is focused on assigning causal blame for the problem and/or offering solutions that prevent gentrification from occurring, or, at the very least, compensating those negatively affected by it. For example, the just green enough (JGE) literature focuses on concerns in communities about potential gentrification and displacement that occur after environmental cleanup. The scholarship offers a preventative approach rooted in local politics. Residents of these communities will still desire environmental improvement, but they may act collectively to ensure that they get *some* environmental improvement, but perhaps not as much as there could possibly be, or perhaps limiting the improvement to a less-desirable level.

We see efforts to prevent gentrification as problematic because cities might avoid providing needed improvements for fear of displacement—even when there is no guarantee that displacement will occur. At minimum, this represents an opportunity cost, as improvements that could be made are not, but at worst this could lead to a choice that *definitely* degrades the quality of life for people (leaving environmental hazards in place) based upon a fear that policy will foster a change that could *potentially* lead to the costs associated with displacement.

With regards to the investment to improve built environments, there are multiple different outcomes to consider, and these outcomes play out over different time horizons. If we consider just health impacts when we make policy choices, then we should change the environmental conditions, but if we consider the risk of displacement in addition to health impacts, then we should be more wary about doing so. Taking action is a short-term strategy that carries longer- term risk, while inaction is a strategy intended to reduce longer-term risk of displacement, but the long-term is, by definition, more uncertain than the short-term. The neighborhood might gentrify in the long-term anyway (see Chap. 6). And, we know that some neighborhoods are suffering from disproportionate pollution absent efforts to improve the neighborhood. Although we have framed this in environmental terms, the same idea could be applied to other policy choices that could affect communities, like economic development programs, or school quality improvements—all offer the potential for short-term community benefits, while risking future gentrification and displacement—and inaction in any of these cases at the very least does nothing to improve the community in the short term, often thus de facto supporting existing injustice.

The frustrating aspect of this for policymakers is that policy can only influence the evolution of the Complex Adaptive System (CAS), it cannot direct it. The system will evolve in the case of either action or inaction, and it will adapt in ways that we may or may not foresee. This is relevant to our fundamental critique of the gentrification literature—gentrification and displacement almost certainly cannot be prevented, but they are also not certain outcomes. Further, we may be able to influence the duration of these processes. Policy changes may not necessarily foster the process, and the process may occur without policy that spurs it, too—and it may even occur regardless of policy efforts to keep it from happening. We see the gentrification literature, and to an extent the politics surrounding the issue, as overly focused on an impossible effort to control an uncontrollable process. Neighborhood change can potentially be influenced, but it almost certainly cannot be stopped, even if stopping change is a worthwhile policy goal, which is questionable.

Policy Vantage Points

We do not want to avoid policy changes that could genuinely benefit people living in poor environmental conditions, yet we do not want to make disruptive changes that drastically change the community in the short term, like worsening environmental conditions or entering a period of rapid gentrification that displaces people with little warning. Policy cannot prevent neighborhoods from changing, especially in the long-term, but ideally it can foster a slower pace of change that enables residents to adapt to those changes. Adaptation in this case might mean slowing the process of gentrification down enough that families are able to achieve stability long enough for children to grow in a healthier community and family wealth to accrue through ownership. Slowing these processes might keep rents from rising dramatically in the short term and thus provide renters with enough time to grow in their careers to begin building wealth so that rental increases do not outpace their income growth. It might allow communities time to foster social ties between members of the community to ensure that wealth increases that occur in the community benefit the existing community rather than only benefitting incoming residents.

It is worth noting, however, that, as this process plays out and wealth is built, the community will change, even if there is no displacement that occurs. But not everyone will benefit. For everyone to benefit is an impossible goal, as some community members will be more vulnerable than others to the risk of displacement even if there is not wholesale displacement. Further, in the converse case, if policy changes are not made, this does not make the community static either, and not everyone will benefit from the lack of change; community members who are more susceptible to pollution, for example, suffer disproportionately from doing nothing. In this case, where fear of displacement stymies change, environmental conditions will not improve, and wealth likely will not be built. The community may be relatively stable, but it might be stable under environmentally degraded conditions that are unhealthy for the members of the community. In this case, some members of the community will build wealth regardless of the circumstances and likely leave the community, while those that remain may end up worse off as property values and other living conditions suffer. In one case, we have displacement occurring, and in the other we have people abandoning the community. Neither of these outcomes is ideal, but they are both going to happen in some communities across a city regardless of what policies are implemented. We see it as a viable goal for cities to take steps to increase the chances that these processes play out in timelines that reduce harm. We see this as much more reasonable than attempting to thwart or prevent changes from occurring.

Conclusion

Environmental improvements in low-income and polluted neighborhoods, where amenities are also lacking, are necessary and welcomed. However, the idea that investments in the built environment may displace those who are supposed to reap the benefits could concern communities and prompt them to oppose such investments that have the potential to ultimately enhance the well-being of residents who have disproportionately endured environmental hazards for a long time. If this understanding results from a lack of clarification in concepts, then a theoretical issue becomes a baseless practical concern rather than helping communities and cities approach the problem constructively. This chapter serves as a reminder of the unintended harm that can result from using the term "gentrification" without clarifying that gentrification and displacement are not the same thing; gentrification increases the risk of displacement, but does not determine it.

If displacement can be considered and examined as separate from–though related to—gentrification, it can be comprehended as an occurrence with different dimensions. Displacement in a gentrifying or gentrified neighborhood might be contingent upon different facets of gentrification, which offers an opening to reevaluate policies capable of exploiting these facets. As discussed here and in other chapters, the frequency, rate, and magnitude of resident displacement can vary by location and neighborhood conditions. We concur with the caution by Elliott-Cooper et al. (2020): "Diagnosing gentrification-led displacement thus requires an attentiveness to its temporal, social and spatial unevenness, and its pernicious impacts on health, quality of life, and well-being" (p. 503).

Here we elucidate the gentrification process while conceptualizing displacement as a distinct process that may arise unintentionally due to gentrification. Conceptualizing gentrification and displacement as different components of related processes broadens our understanding of gentrification and enables us to view displacement from a different angle. Instead of assuming the potential negative effects of governmental policies aimed at improving the environment in low-income and polluted areas, it is important carefully to consider the circumstances under which these efforts may or may not result in significant displacement.

References

Anguelovski, I., Connolly, J. J. T., Cole, H., Garcia-Lamarca, M., Triguero-Mas, M., Baró, F., Martin, N., Conesa, D., Shokry, G., Del Pulgar, C. P., Ramos, L. A., Matheney, A., Gallez, E., Oscilowicz, E., Máñez, J. L., Sarzo, B., Beltrán, M. A., & Minaya, J. M. (2022). Green gentrification in European and North American cities. *Nature Communications, 13*(1), 3816. https://doi.org/10.1038/s41467-022-31572-1

Chapple, K., Waddell, P., Chatman, D., Zuk, M., Loukaitou-Sideris, A., Ong, P., et al. (2017). *Developing a new methodology for analyzing potential displacement*. UC Berkeley. https://escholarship.org/uc/item/6xb465cq

Eckerd, A., & Reames, T. (2012). Urban renaissance or invasion: Planning the development of a simulation model of gentrification. In *Simulation for policy inquiry* (pp. 83–100). Springer.
Eckerd, A., Kim, Y., & Campbell, H. (2019). Gentrification and displacement: Modeling a complex urban process. *Housing Policy Debate, 29*(2), 273–295. https://doi.org/10.1080/10511482.2018.1512512
Ellen, I. G., & O'Regan, K. M. (2011). How low income neighborhoods change: Entry, exit, and enhancement. *Regional Science and Urban Economics, 41*, 89–97.
Elliott-Cooper, A., Hubbard, P., & Lees, L. (2020). Moving beyond Marcuse: Gentrification, displacement and the violence of un-homing. *Progress in Human Geography, 44*(3), 492–509. https://doi.org/10.1177/0309132519830511
Freeman, L. (2005). Displacement or succession? Residential mobility in gentrifying neighborhoods. *Urban Affairs Review, 40*(4), 463–491.
Gould, K. A., & Lewis, T. L. (2017). *Green gentrification: Urban sustainability and the struggle for environmental justice*. Routledge: London and New York.
Hamnett, C. (1991). The blind men and the elephant: The explanation of gentrification. *Transactions of the Institute of British Geographers, 16*(2), 173–189.
Hamnett, C. (2003). Gentrification and the middle-class remaking of inner London, 1961-2001. *Urban Studies, 40*(12), 2401–2426. https://doi.org/10.1080/0042098032000136138
Hwang, J., & Lin, J. (2016). *What have we learned about the causes of recent gentrification?* (Working paper pp. 16–20). Federal Reserve Bank of Philadelphia. https://doi.org/10.21799/frbp.wp.2016.20.
Ilic, L., Sawada, M., & Zarzelli, A. (2019). Deep mapping gentrification in a large Canadian city using deep learning and Google Street View. *PLoS One, 14*(3), e0212814.
Ley, D. (1986). Alternative explanations for inner-city gentrification: A Canadian assessment. *Annals of the Association of American Geographers, 76*(4), 521–535.
Lyons, M. (1996). Gentrification, socioeconomic change, and the geography of displacement. *Journal of Urban Affairs, 18*(1), 39–62. https://doi.org/10.1111/j.1467-9906.1996.tb00365.x
Marcuse, P. (1985). Gentrification, abandonment, and displacement: Connections, causes, and policy responses in New York City. *Urban Law Annual. Journal of Urban and Contemporary Law, 28*. https://openscholarship.wustl.edu/law_urbanlaw/vol28/iss1/4
Naik, N., Philipoom, J., Raskar, R., & Hidalgo, C. (2014). Streetscore – Predicting the Perceived Safety of One MillionStreetscapes, 2014 *IEEE Conference on Computer Vision and Pattern Recognition Workshops*, Columbus, OH,USA, pp. 793–799, https://doi.org/10.1109/CVPRW.2014.121
Naik, N., Kominers, S., Raskar, R., Glaeser, E., and Hidalgo, C. (2015). *Do people shape cities, or do cities shape people? The co-evolution of physical, social, and economic change in five major U.S. cities* (National Bureau of Economic Research Working Paper 21620).
Richardson, J., Mitchell, B., & Franco, J. (2019). *Shifting neighborhoods: Gentrification and cultural displacement in American cities*. National Community Reinvestment Coalition. https://ncrc.org/gentrification/
Shokry, G., Connolly, J. J. T., & Anguelovski, I. (2020). Understanding climate gentrification and shifting landscapes of protection and vulnerability in green resilient Philadelphia. *Urban Climate, 31*, 100539. https://doi.org/10.1016/j.uclim.2019.100539
Smith, N. (1987). Gentrification and the rent gap. *Annals of the Association of American Geographers, 77*, 462–465.
Stuhlmacher, M., Kim, Y., & Kim, J. E. (2022). The role of green space in Chicago's gentrification. *Urban Forestry & Urban Greening, 71*, 127569. https://doi.org/10.1016/j.ufug.2022.127569
Zuk, M., Bierbaum, A. H., Chapple, K., Gorska, K., & Loukaitou-Sideris, A. (2018). Gentrification, displacement, and the role of public investment. *Journal of Planning Literature, 33*(1), 31–44. https://doi.org/10.1177/0885412217716439

Chapter 5
New Amenities that Provide Jobs: Consequences of Density and Segregation

Abstract To shed some light on longstanding questions around gentrification, in this research we model environmental gentrification and gentrification-related displacement of residents. We do this through the development of an Agent-Based Model (ABM) of a simple urban region, considering different urban contexts and policy approaches to polluted facilities and the relationship of these policies with subsequent gentrification and displacement. We find that gentrification-related displacement is most likely, and most impactful, in urban regions characterized by high levels of density and low levels of residential segregation preferences. Displacement is far less prevalent in low-density regions, particularly those with high segregation preferences. We discuss the potential for different policy implications in these different urban contexts.

Introduction

Gentrification is a difficult concept to define, to study, and to craft policies for (Rayle, 2015). The term is ideologically laden (Freeman, 2011), it is difficult to conceptually separate and measure the various aspects of the gentrification process (Slater, 2006), and the line between economic development and gentrification is sufficiently blurred that crafting policies to address the issue is a challenge (Clark, 2005).

This chapter is a reprint of the following article: Eckerd, Adam, Yushim Kim and Heather E. Campbell (2019). "Gentrification and Displacement: Modeling a Complex Urban Process." *Housing Policy Debate* 29(2): 273–295. Published online 26 November 2018. DOI: https://doi.org/10.1080/10511482.2018.1512512. It is reprinted by permission of Informa UK Limited, trading as Taylor & Francis Group, © Virginia Polytechnic Institute and State University, www.tandfonline.com on behalf of Virginia Polytechnic Institute and State University. Minimal changes have been made in order better to fit the chapter's location in this book; this includes changes to figure numbers and reference style (etc.) and the addition of some footnotes, so footnote numbers will not be identical across the article and the chapter. Insertions are also indicated by the use of square brackets.

H. E. Campbell et al., *Green Gentrification and Environmental Injustice*,
https://doi.org/10.1007/978-3-031-65100-7_5

In general terms, gentrification is thought to occur when formerly lower socioeconomic status neighborhoods transition to higher socioeconomic status (Hamnett, 1991). This implies a two-part process—the movement in of higher socioeconomic status residents, and the moving out of lower socioeconomic status (SES) residents (Slater, 2006). This inflow and outflow of residents can be considered part of the same process, as Glass (1964) conceived of gentrification when she coined the term, but they may also be fundamentally different processes. Further, it is the flows out of lower-SES residents that are considered to be potentially problematic, for these may be involuntary and harmful to lower-SES residents; but an inflow of higher-SES residents that does not displace lower-SES residents may have economic [and social] benefits. Our approach in this [chapter] is to disentangle this complex issue through the use of a simulation model that allows us to explore the in-migration of high-SES residents separately from the out-migration of lower-SES residents[, and to focus specifically on the effects of adding amenities that provide jobs and on the levels of segregation].

Specifically, we consider the nature of the relationship between the in-migration process, which we will refer to as *gentrification*, and the out-migration process, which we will refer to as *displacement*, or gentrification-related displacement. In doing so, we address challenges of assessing and measuring gentrification and displacement (Grossmann & Haase, 2016) through the use of an agent-based simulation model (ABM) of an urban system that enables us to track the movement of city residents over time. Although the model is a simulation of a city, it allows us to explore the possibilities for and pressures on different types of residents as a city grows and changes over time. We track both where gentrification and displacement take place, and also the burdens that are placed on residents in neighborhoods that gentrify, specifically assessing the extent to which displacement of existing residents occurs and what conditions exacerbate or ameliorate this outcome (Eckerd & Reames, 2012) [, with a particular focus on segregation].

In the sections that follow, we describe challenges of studying gentrification and displacement, how gentrification is conceived in the literature, and then describe our approach and model. The simulation results follow, and we end with some implications for housing policy.

Gentrification

Gentrification has been defined in a variety of different ways, ranging from benign processes of housing market sorting (Smith, 1979), to militaristic metaphors of invasion and conquest (Hamnett, 1991). At root, gentrification is a process of neighborhood change (Grossmann & Haase, 2016). To gentrify, a neighborhood is first populated primarily by lower wealth residents and subsequently populated by higher wealth residents. While some view the process as a fundamentally class-based phenomenon (Glass, 1964), others call attention to racial components as white populations increase in neighborhoods that had previously been primarily

African American or Hispanic (at least in the American context) (Schaffer & Smith, 1986; Hwang & Sampson, 2014). Some argue that gentrification only occurs if lower wealth residents are displaced (LeGates & Hartman, 1986), whereas others contend that gentrification and displacement are conceptually distinct (Slater, 2006).

Here we take the latter perspective, for there are wide-ranging implications of gentrification and, despite the general tone of much of the literature, not all are necessarily negative (Freeman, 2008). Although reinvestment may displace people, it can also bring more jobs, services, and safety to neighborhoods that have lacked these needs (Bartik, 1991; Valler & Wood, 2010). Some residents can benefit financially through higher property values and deconcentration of poverty (Sullivan, 2006). Of course there are also possible negative implications, too. Residents may well lose their sense of what their neighborhood was and who it was for (Freeman, 2008), or end up worse off than they were before if rents rise and they are forced to move into less appealing neighborhoods and also bear the burdens of moving (Vigdor, 2002).

If reinvestment benefits both existing and incoming residents, we may want policy to encourage more redevelopment, but if existing residents are burdened with displacement, then gentrification may need to be tempered (Clark, 2005). That is, if little displacement occurs, the concept of gentrification may not be fundamentally different from economic development (Sullivan, 2007), but if much displacement occurs, then the implications of this process could be burdening some of the very people that economic development is usually intended to benefit.

This is all complicated by the fact that we do not know the scale of the displacement problem (Slater, 2006). We certainly have evidence that displacement occurs in certain areas (Lyons, 1996; Vigdor, 2002), but there is little generalizable evidence regarding just how much displacement occurs (Hwang, 2016).[1] Schill and Nathan (1983) and Vigdor (2002) note that displacement, while a key factor in considerations about the implications of gentrification, can be very difficult to assess. It is problematic to differentiate between residents who were displaced from those who would have relocated anyway. This may be because we only know that gentrification has taken place after the fact, or at least once the process begins. This means that we cannot readily identify areas that *will* gentrify, only those that either *have* or are currently undergoing gentrification, and so potential displacees cannot be identified and tracked prior to the gentrification. This is particularly problematic because the first displacees may well be the only people displaced, and are the most vulnerable assuming that the first movers are those least able to afford rent increases (Ley, 1981; Clay, 1979; Smith, 1979; Fusch, 1980; Hammel & Wyly, 1996; Wyly & Hammel, 1998; Kennedy & Leonard, 2001).

Tracking the movement of individual residents is difficult to do on a scale large enough to provide generalizable insight—it is easier to assess who has moved into a geographic area than to assess the disparate movements of people away from that

[1] [Since this article was published, an NCRC report does provide some: Richardson, J., Mitchell, B., & Franco, J. (2019), *Shifting neighborhoods: Gentrification and cultural displacement in American cities*. www.ncrc.org. Significant discussion of this is presented in other chapters.]

area (Vigdor, 2002). Further, even if residential movement could be tracked systematically, it would be difficult to separate the circumstances under which individuals left gentrifying communities. Some may have been displaced because of higher rents, but others may have relocated for work, others might have moved regardless of gentrification, while others still may have been taking advantage of increases in their home's value to sell in a rising real estate market (Rayle, 2015).

Further, gentrification processes take place in the broader context of urban residential sorting patterns (Sanchez & Dawkins, 2001), and are affected by land uses (Wyly & Hammel, 1998), and economic considerations like commuting costs (Brueckner & Rosenthal, 2009)—themselves affected by the transportation infrastructure. They are affected by the housing stock (Smith, 1979) and living preferences of the regional population (Ley, 1986) [, with one relevant living preference being the level of segregation]. [So,] identifying when gentrification and gentrification-related displacement occur is challenging [and needs to consider a mixture of variables].

Some types of cities appear to be more prone to gentrification than others. Large cities, especially those with rapid job growth and tight housing markets, seem more prone to gentrification, as do cities with high traffic congestion (e.g., inefficient transportation infrastructure), and public policies (notably tax breaks) that encourage people to live in redeveloping neighborhoods (Black, 1980; Rose, 1984; Wyly & Hammel, 1998; Kennedy & Leonard, 2001). Gentrification may be more prone to occur in cities with both high transportation costs, but also in those with good access to public transit. It may also be tied to the design of the urban setting, in particular the architecture of available buildings and the aesthetic nature of the community (Kennedy & Leonard, 2001; Kolko, 2007; Eckerd, 2011). Gentrification may also be more likely in areas where residents prefer, or at least accept, higher levels of density—sometimes due to geographic features, general residential preferences, or even policies intended to curb sprawl (Dieleman & Wegener, 2004).

Other public policies have been used as economic development drivers that have also been tied to gentrification. One hallmark of recent urban redevelopment is the remediation of environmental hazards, transforming polluting disamenities into new urban amenities, like parks, mixed-use development projects, or creative spaces. In certain locations, these efforts have been tied to subsequent gentrification (Checker, 2011). This transition from disamenties to amenities can create incentives for investment in a community that would not have previously appealed to outsiders due to the disamenity (Eckerd, 2011). However, this process is difficult to observe as well, because not all regions that receive improvement will gentrify.

In short, there are complex, interweaving systems of urban policy, community activism, residential taste [(including segregatory tastes)], transportation systems, and real estate market dynamics that foster, but do not promise, gentrification in specific locations. Empirically, it is difficult to know when and how gentrification may occur, if and which residents will be displaced, and the implications of this displacement. Additionally, there are challenges because studies show that different ways of measuring gentrification, in terms of demographic outcomes, the changes in amenities, and the areal unit of analysis one uses, often lead to different conclusions (Eckerd, 2011).

When we consider the many factors and interactions that go into determining whether gentrification occurs and its consequences for displacement, it becomes apparent that gentrification and displacement occur as emergent outcomes of a complex system. A complex system is characterized by a set of interacting components, such as economic systems or social systems, where the aggregate result of these interactions is a nonlinear, emergent response (Miller & Page, 2007; Holland, 1998). In simpler terms, the emergent occurrences of a complex system are different and more than the sum of its parts. When complexity is a key feature of system dynamics, then simulation methods, such as Agent-Based Modeling, can be a useful approach to further understanding (Epstein, 2006). Rather than aiming to differentiate the independent effects of some set of independent variables on some outcome variable, simulation methods help observe the behavior of a system that emerges from interdependent inner-working mechanisms. Therefore, the results of any particular simulation trial are the consequence of an interacting set of individual actors within a dynamic setting. The assumption is not that there are independent effects that lead to some outcome, but rather that systemic social patterns emerge from a distinct *recipe* of the interaction of different social systems. Other scholars have also used computational simulation methods to understand related issues (Epstein, 2006).

The utility of ABMs for urban analysis is well established (Batty, 2007). ABMs have been used to learn more about urban development and sprawl (Brown & Robinson, 2006), emergency management strategies (Chen & Zhan, 2008), environmental justice (Campbell et al., 2015), and land use (Filatova et al., 2009). Complexity analysis techniques such as computational simulations have been used in studies of traffic (Nagel & Rickert, 2001), land-use (Engelen et al., 1995), and settlement patterns (Dean et al., 2000). Beauregard (1986) was perhaps the first to explicitly connect the notions of complexity and gentrification, and since then several models have assessed the issue. O'Sullivan's (2002) model suggests that, as real estate value decline reaches its nadir, property values in neighborhoods with amenities can suddenly and rapidly increase, such that a neighborhood can fully gentrify within just a few years. Torrens and Nara (2007) find that gentrifying neighborhoods may tend to remain distinct from other nearby areas, suggesting that the geographic limits of gentrification may be relatively small, perhaps defined by amenities, and that gentrification can be rather sporadic. Other related models aim to unpack the individual-level location-choice decision (Devisch et al., 2009), assess individuals' propensity to consider the local policy context (O'Sullivan, 2009), consider how urban spatial constraints and economic pressures affect residential patterns (Haase et al., 2010), or how land-use change affects cities (Huang et al., 2014) or specific regions (Jackson et al., 2008). Broadly speaking, this work, like much of the empirical work on gentrification, has tended to assess the systemic outcome of decision making processes of incoming residents but neither the scale of, nor the burden placed on, displaced residents.

Our model is an attempt to broadly investigate gentrification and displacement. We fold in aspects of the things we know are associated with gentrification—proximity to amenities (and distance from disamenities), access to employment, an urban setting that tends to have more density than a suburban setting, and living in proximity to people similar to oneself (Kennedy & Leonard, 2001; Helms, 2003; Kolko, 2007; Eckerd, 2011)—and explore the potential for and the implications of displacement of the original residents.

The Agent-Based Model

As the above discussion underscores, gentrification and displacement are complex phenomena. They are not easily understood or predicted using conventional policy analysis, and are challenging to assess quantitatively, and while case studies have value, it is difficult to know if the trends identified in particular cases provide a generalized understanding of the phenomenon. To address these challenges, we model this process using an agent-based computer simulation model, based on a real estate market across a set of different regional contexts.

Our approach with the model is parsimonious—meaning that we do not try to explicitly model the complexity of the real world, but rather to simplify a real urban system into a few key parts. We do so purposely in order to have a model that represents the complexity of a true urban system, but is also simple enough to draw meaningful conclusions. The model is described in detail below, but, stated simply, the model is of an urban region consisting of heterogeneous residents with different characteristics such as race (majority or minority); wealth (drawn randomly from empirically based distributions); and living preferences who select locations in a housing market affected by amenities and disamenities, residential movement, and density, quality, and price. Therefore, while some of the known factors associated with gentrification and displacement are explicitly contained in the model, some other factors noted above as important to gentrification and displacement are more implicitly contained in this structure. For example, a density preference could be thought of as a corollary to the transportation infrastructure, since we would expect residents to exhibit a greater preference for density (cet. par.) when transportation is less efficient.

One advantage of the ABM is that we can track the movement of all of the agents throughout the entire simulation. We trace residential movement through time, noting where neighborhoods gentrify, how many residents are displaced due to gentrification, where residents who leave gentrified neighborhoods go, and how many times residents in certain demographic groups are displaced.

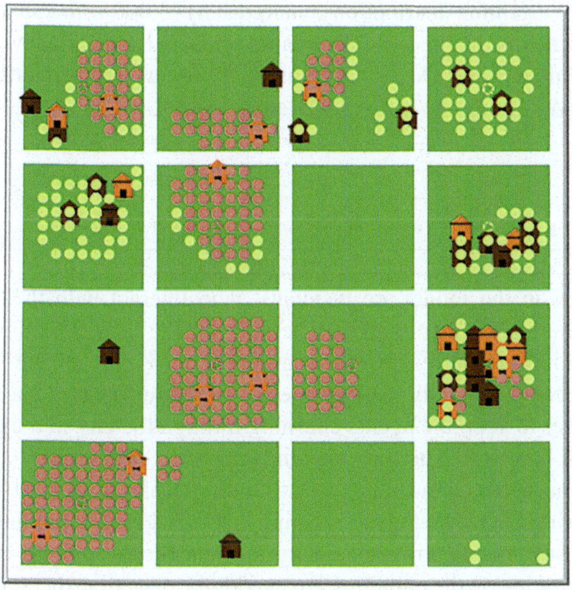

Fig. 5.1 Graphic depiction of the urban region partway through one scenario
Note: Dots represent residents (e.g., majorities or minorities. Houses represent land uses (e.g., disamenities or amenities).

Model Structure

A graphical depiction of the model is provided in Fig. 5.1. The urban region is modeled as a simple grid of 41 × 41 (1681 total) plots. These plots are the locations that residents can choose as their *home* plots. In other words, the plots can be conceived of as land locations that residents (and *firms*) select. Some plots will be occupied and some vacant and thus open to new (residential or non-residential) land uses, depending on the characteri[sti]cs of the setting, as described below. In some cases, modeling higher density preferences by residents, more than one *house* can be placed on a plot.[2] The size of the artificial city was calibrated to ensure that there was sufficient space for a large population of residents, but also relatively confined enough to limit residential choice and to ensure that the simulation could compute efficiently on our available computing resources. In the simulation trials described below, we alter many of the characteristics of this *city* but we do not alter its geography from this grid. Plots have variable prices and quality measures, both of which affect the desirability of each plot to the residents. Price and quality are adjusted as described below. At the start of the simulation, the expected quality variable is set to 50 for all plots (and can ultimately range from 0 (the lowest quality) to 100 (the highest quality) as the simulation moves forward in time). Prices are drawn

[2] When residents have higher density prefer[e]nces, modeled cities have higher density housing outcomes.

randomly from a gamma distribution centered at $173,000 with a standard deviation of $43,000, which is the median home value price and standard deviation for tracts in the 2010 United States Decennial Census. We use a gamma distribution because it is right-skewed (as is true of housing values in the US).

Residents

The key actors in the model are residents, who interact with their environment, which includes one another and characteristics of the land. The simulation begins with a population of 200 residents that grows at a 5% rate at each tick forward in time—the process is actually to eliminate 2% of the residents at each time tick and add 7% in order to keep the agents churning through the city in a dynamic process as time moves forward.[3] The time encompassed by a *tick* is fairly arbitrary, but in this context can be thought of as 1 year. Unlike Haase et al. (2010), we model a growing city since a city that is not growing is less likely to be adding residents and experiencing the housing shifts necessary for gentrification. Land Uses (LUs) that are either amenities or disamenities are randomly placed at the start (see below), with new LUs added at the rate of one each per every 10 new residents.

Residential Attributes (Kind and Wealth)

Residents are intended to be simple but heterogeneous representations of people. They function according to only one goal—find a location to live—and have two key attributes: *kind* and *wealth*. Kind is a variation on ethnicity or race. We use kind rather than these more familiar terms because our model does not include any historical discriminatory behavior or social distinction—kind represents a simple, mathematical majority/minority distinction, rather than a more pervasive minoritization process that characterizes racial and ethnic social relations. However, some discrimination can be included through the similarity preference function, discussed below. At the start of the model, 70% of the population is majority and 30% is minority. The population growth function randomly replicates the kind proportion of the population; thus, while the proportional breakdown can vary from this exact state, the proportions remain relatively consistent around this point.

To mirror the American context of our study, we assume that the majority is wealthier, on average, than the minority. Each time a new resident enters the model, it draws a wealth value from one of two gamma distributions, with the majority income distribution centered at $54,000 (the median family income of White Americans in 2010) and the minority income distribution centered at $32,000 (the

[3] These eliminations can be thought of as people moving completely out of this city to another, or also as losses because of death, etc.

median family income of African Americans in 2010), with both distributions defined by a standard deviation of $40,000 (Campbell et al., 2015).[4] As the simulation moves forward in time, residents' wealth levels change based on real estate transactions. This is done to account for the potential monetary benefit that homeowners in gentrifying areas may see when they sell their home for more than they paid for it. When residents move, their wealth is increased by the difference between what they *paid* for the home and what they *sold* the home for (or vice versa, wealth can decrease if they *sell* for less than they *paid*).

Resident[i]al Location Decision-Making

New residents, including those present at the initialization of the model, or existing residents that are considering relocating, select a location using a basic utility function, subject to two constraints. Residents begin the location process with a set of plots that are available (i.e., those that are not roads and do not contain non-residential LUs). Residents first satisfy the two constraints and then locate based on the plot of land that provides them with the highest utility within their constrained set of plots. For the first constraint, we assume that residents will not locate to a plot that they cannot afford, nor to one that is substantially cheaper than they can afford. We operationalize this in the model by assuming that residents will not select a plot that is more than 3 times greater than their individual wealth level, a common recommendation in real estate purchases, nor a plot that is less than their wealth. If no plots are available that fit this constraint, residents will remain at their current location.

For the second constraint, residents are also constrained by a similarity preference—which we model as an implementation of Schelling's (1971) insights, insights that have also been shown to be important when considering environmental justice outcomes (Campbell et al., 2015). In line with Schelling, we assume that residents will limit their choice-set of locations to ensure that some proportion (varied in each scenario) of other residents near them is of the same kind as themselves, and (as Schelling did) we model this as a constraint on location choice. The similarity preference can be thought of as an articulation of the level of discrimination via housing choice. Given previous research illustrating how important this similarity preference may be for location choices (Campbell et al., 2015), we also vary the similarity preference scenarios as described below. Higher or lower similarity preferences lead to cities that are more or less segregated.

In the model, the similarity preference functions as a constraint—residents will only calculate utility scores of plots that fit their similarity preferences (of having [at

[4] The gamma distribution is used because it is right-skewed, as is the US income distribution. Both distributions are constrained to be no less than zero. Data for the income variable was derived by the authors from the 2010 US Census. Using the Census block level as the unit of analysis, we imputed a dataset of individuals in the US via the population of each Census block, and a random assignment of individual incomes based on the average income in each Census block.

least] 30%, 50%, or 80% of the residents on contiguous plots be of the same kind as themselves). If no plots are available that fit this constraint, residents will either remain at their current location, or, if they are new to the region, they will relax the similarity preference and select a plot that fits their price constraint and maximizes their utility.

Once these constraints are satisfied residents select the plot in their constrained set that provides them with the highest individual utility. The utility function is based on how desirable any particular plot in the region is to each heterogeneous agent. The utility function balances each agent's distinct preferences with respect to price, quality, distance to amenities and disamenities, access to transportation routes, and density as arguments.[5] All preferences are unique to each individual agent—each agent is randomly assigned a price, quality, distance to LUs, access to transportation, and density preference from a normal distribution. That is, some residents may have a high concern about price, while others may be less concerned about price and more concerned about transportation access. The price, quality, distance to LUs, and access to transportation distributions are the same for all simulations. In some cases we vary the mean for the density preference distribution to simulate different types of cities, at means of 1 resident per plot, or 5 or 10 residents per plot. Thus, each agent's individual location preferences are variable, but still clustered around a common point.

When residents select a location, they consider their individual preferences and compare each plot to see which one maximizes their preference structure. The utility function is shown in Eq. (5.1):

$$u_{i,t} = p_{i,t}^{-\alpha} \cdot q_{i,t}^{\beta} \cdot l_{ik,t}^{-\gamma} \cdot r_{i,t}^{-\delta} \cdot d_{i,t}^{\varepsilon} \tag{5.1}$$

An individual agent's utility rate, u, for plot i at time t is a function of the price, p, of plot i at time t; the environmental quality, q; the shortest distance, l, to some LU, k; the shortest distance, r, to a transportation route; and the density value, d. In this function, we assume that residents would prefer low prices, high quality, to minimize distances to either amenity or disamenity LUs (we assume that all LUs offer some benefit to the resident, such as a place to work or spend leisure time), to minimize distances to transportation routes to minimize commute times, and a density level that is near their individual preference. The parameters α, β, γ, δ, and ε are weighted for each individual resident, assuming that residents have different LU preferences (Torrens & Nara, 2007).

After these constraints have been satisfied, all newly introduced residents move to the plot that has the highest utility score in their available choice set. *Existing* residents are mobile and may or may not relocate depending on a comparison between their current home and other plots in the region. As the simulation steps

[5] We thank an anonymous reviewer for suggestions about the utility function and the modeling strategy generally. After conducting sensitivity analyses, the model was altered to better capture the heterogeneity of agent decision making. See the Appendix for more detail about the model.

forward in time until the end (tick = 60), at each time tick 20% of the existing residents are randomly selected to consider relocating (Campbell et al., 2015). Those selected seek a location that fits their price and similarity constraints and scores the highest utility value as described above. If they find a location with a higher utility value than their current location, they relocate. If there are no plots in their constrained choice set that increase utility above their current location, they do not move, except for under one exception. If the price of their current plot sufficiently exceeds their current wealth (as described in more detail below), then residents will move and we consider them to be *displaced*.

Displacement

Displacement in this sense can be conceived of in one of two ways; either home-owning residents are compelled to sell their home when its value becomes quite high relative to their wealth (e.g., because they can no longer afford property taxes), or, if residents are conceived of as renters, the rent exceeds their ability to pay and they are forced to relocate. We note that the consequences of these two options are in reality different since when owners sell they increase their wealth, but renters do not. As described below, we consider either description to illustrate possible displacement, even if the former might be considered utility-enhancing while the latter is clearly utility-reducing. Residents are displaced whenever the price of a plot is at least double the resident's current wealth. When this is the case, the resident relocates to a different plot that fits their price and similarity constraints. The proportion of residents that actually does move at each tick forward varies as space becomes constrained by a growing population over time, so in early time periods the proportion of residents that relocates is close to 20%, while in later time periods this decreases to about 10%. This seems realistic, as populations tend to be more prone to relocate whenever there is ample space and some perceived risk of remaining in place (Kirschenbaum, 1996).[6] In the model, as the city matures, the quality levels stabilize, reducing the incentive of residents to relocate. Through the trials, displacement occurs at lower rates than the rate of random relocation, but, as shown below, is prevalent.

Land Uses

LUs can be conceptualized as any non-residential land use, and we distinguish between amenities and disamenities in a general sense. Amenities might be non-polluting firms, parks, schools, commercial districts, or other LUs that residents

[6] [In their empirical analysis of New York City's Housing and Vacancy Survey from 1991 to 2002, Newman & Wyly (July 2006) find displacement rates between 6% and 10% per year (p. 3).]

might prefer to locate near. Disamenities might be polluting industries, polluted plots of land, transportation centers, jails, or any other LU that residents would prefer to live away from (if they don't work there) but that still offer some potential value. LUs are randomly assigned an amenity score ranging from 0 to 9. A score of 0 is a pure amenity, and 9 is a pure disamenity, with scores of 0–4 indicating that an LU is on balance an amenity and 5–9 a disamenity. The lower the amenity score, the more residents seek to be located nearby. These amenity score levels are drawn from a uniform distribution. Thus, our assumption is that there are comparatively few pure amenities or disamenities, but rather a set of LUs that offer some benefits and some detriments. As mentioned above, LUs are added to the world at the rate of one for every ten new residents. LUs select locations based on finding a plot that is not inhabited by any residents and minimizing price.

Land

Plots of land are modeled to function according to a real estate market, calibrated to function as realistically as possible given the model's parsimonious design.

Plot Price

Each plot's price is affected by residential location decisions nearby and the presence of LUs. First, plot prices are decreased based on the inverse distance to LUs. Basically, pure amenities with a score of 0 have no effect on plot price, but all other LUs have some negative effect on price—which may be indirectly balanced out by residential demand and resident preferences to locate near amenities. Plot prices are also weighted at each tick by the average wealth of the residents located on the plot, and those on the eight contiguous plots (Veldkamp & Verburg, 2004), as shown in Eq. (5.2):

$$p_{m,t} = \left[p_{m,t-1} - \left(p_{m,t-1} \left(0.01 * s_{k,t} \right) \right) \right] + \left[\frac{\sum w_{i,t}}{n_i} + \frac{\sum w_{j,t}}{n_j} - p_{m,t-1} \right] \tag{5.2}$$

In Eq. (5.2), the price, p, of plot m at time t is a function of its price at time $t - 1$, decreased by 1% of the amenity score, s, of the nearest LU,[7] k, at time t, balanced by the average wealth, w, of the residents, i, located on the plot and the average wealth of the residents, j, located on the eight contiguous plots (recall that more than one *house* can be on a plot under some density conditions). Basically, plots with wealthier residents nearby, and closer to amenities cost more (and vice versa).

[7] [Recall that LUs with high values are disamenities, and with low values are amenities.]

Plot Quality

Quality is adjusted similarly, although this value is not affected by residents and is only affected directly by the closest LU. Quality is adjusted up or down based on the amenity score of that nearby LU, as shown in Eqs. (5.3a and 5.3b):

$$\textit{If } k = 0-4 : q_{m,t} = \left[q_{m,t-1} + \left(q_{m,t-1} \left(0.01 * \left(4 - s_{k,t} \right) \right) \right) \right] \tag{5.3a}$$

$$\textit{If } k = 5-9 : q_{m,t} = \left[q_{m,t-1} - \left(q_{m,t-1} \left(0.01 * s_{k,t} \right) \right) \right] \tag{5.3b}$$

In these equations the quality q for plot m at time t is increased as in Eq. (5.3a), or decreased based on Eq. (5.3b), depending on if the closest LU, k, is an amenity or disamenity. Quality is then altered by comparing the focal plot's quality level with the average quality of the eight contiguous plots. If the average quality level of the contiguous plots is greater than the quality level of the focal plot, then the focal plot's quality is increased by 1% of the difference. Conversely, if the focal plot's quality is higher than the average of the contiguous plots, then its quality is decreased by 1% of the difference.

Defining Model Outcomes

Gentrification

The literature on gentrification does not offer clear guidance on when, precisely, a neighborhood becomes gentrified. Without clear empirical guidance, we operationalize it as a calibrated process. First, a plot must be gentrifiable—that is, it must reach a state where it declines enough to become a potential target for gentrification. We conceive of this simply—if a plot's price is in the bottom quartile of plot prices we consider it gentrifiable. We then consider a plot to be gentrified only if it first becomes gentrifiable, and then subsequently sees a price increase into the upper two quartiles of price.[8] Once land is gentrified, it is no longer gentrifiable, unless it subsequently drops back into the lowest quartile of land values.

[8] As a first attempt to operationalize this concept in a simulation model, we arrived at this operationalization through testing. We set both lower thresholds for *gentrifiable* and higher thresholds for *gentrified*, settling on these values as a good balance of ensuring that some gentrification took place in most trials but did not overwhelm the entirety of the region. This operationalization also seems to fit common-sense notions of what it means for a neighborhood to be gentrified.

Displacement

We assess the amount of displacement from gentrified plots. Because residents are mobile and some move for different reasons, we assume that residents can only be displaced if they lived on (a) a gentrifiable plot that was (b) subsequently gentrified while they were still located there, and then (c) relocated to a different plot that was gentrifiable (but not gentrified at that time) because they met our housing burden threshold as described above. We believe this reasonably captures the type of displacement that is of concern to those worried about gentrification-related displacement. We track the residents that move, for this reason, from gentrified plots to gentrifiable plots, assessing these rates for different wealth and kind classes of residents.

Scenarios and Setup

To ensure that we begin with a region that has ample opportunity for gentrification and displacement, we begin each trial by engineering the region to have relatively high levels of segregation and disparities between the kinds of residents. Basically, we try to mirror an American urban setting, around the time we began to note gentrification occurring (mid to late 1980s) and assess results from this starting point. For the first 25 time ticks[9] of each scenario, we ensure that the majority group lives on plots with higher quality levels and that both groups locate with very high similarity preferences (each resident class prefers 80% of residents on the eight contiguous plots to be of the same kind). After tick 25, we alter the decision-making process for residents and begin a phased removal of disamenities in order to spur redevelopment and potential gentrification (Checker, 2005; Eckerd, 2011). At each tick after tick 25, we assess gentrification and displacement in a variety of different contexts and assess what happens to the residents involved.

Disamenity Removal Policy Scenarios

In all scenarios we implement a policy removing one existing disamenity from the region per time tick as our method to spur gentrification. As shown in Table 5.1, in some scenarios the disamenity that is on the highest priced plot is removed (illustrating an assumption that the plot has redevelopment potential), and in others the disamenity located near the largest clusters of minority residents is removed (illustrating a scenario of targeted cleanup of minority areas). We also vary what occurs after the disamenity is removed: in some cases, it is simply removed, while in others

[9] For simplicity, we conceive of each time tick as 1 year, but the time ticks may represent any relevant time designation.

Table 5.1 Scenarios and variable settings

Variables	Distinct assumptions		
Disamenity priority	Clean up disamenity on highest-priced plot	Clean up disamenity on plot with highest density of minority residents nearby	
Disamenity process	Disamenity is removed; land is available for land uses and residents	Disamenity is removed and replaced with an amenity	
Similarity preference	No segregation: 0% for both majority and minority groups	Low segregation: 30% preference for both groups	High segregation: 80% majority preference, 50% minority preference
Density preference	Low density: Mean density preference of 1 resident per plot	Moderate density: Mean density preference of 5 residents per plot	High density: Mean density preference of 10 residents per plot

a disamenity is *redeveloped* into an amenity. We expect that when disamenities are removed the plot attributes will be affected by both an increase in quality and price, which will change how this plot is perceived through the residential utility function. We assume that removal of the disamenity will immediately change quality, akin to removal of a hazardous site, but the price change will lag behind, making the site attractive to new or relocating residents who can take advantage of low prices and comparatively higher quality. We expect both of these processes to quicken when the disamenity is immediately redeveloped into an amenity—quality will rise more quickly, and price will follow closely.

Residential Preference Scenarios

We vary the context in two ways: by changing similarity preferences and density preferences to account for variation in residents and their cities. Since each *city* we compare has the same geographic form, we varied similarity preferences to account for cities that vary in the extent to which there is high or low racial segregation in settlement patterns, and density preferences to account for cities that are generally more or less dense. The density preference could also be thought of as a geographic constraint. In some highly dense cities, residents may simply prefer more dense settlement or there may be geographic constraints (like waterways or mountains) or policy constraints (like urban growth boundaries) that enforce higher density. From our modeling perspective, the result is the same—in some cities residents live in dense urban settings, while in others there is primarily less-dense, more suburban-style settlement. For our scenarios, we varied similarity preferences in three ways—two hypothetical situations (0%, 30%) and an empirical finding (50% minority/80% majority similarity preferences, Clark, 1992)—density preferences in three ways (low, medium, and high), disamenity removal priority in two ways (as described above), and disamenity removal processes in two ways (all as depicted in Table 5.1). This results in a total of 36 distinct scenarios (3 × 3 × 2 × 2). We ran each scenario for 100 trials each with 60 ticks per trial,[10] and aggregated the means of all key measures for the trials for each scenario.

Results

Policy Choices

In our initial analysis of the results, it was clear that some of our experimental variables had very little effect on the outcomes of focus. Whether disamenities were cleaned up or cleaned up and replaced with amenities had virtually no effect on gentrification or displacement outcomes. Further, results were also indistinguishable whether cleanup was targeted to spur economic development by remediating the highest priced plot or to improve environmental equity by remediating disamenites in minority areas. These findings alone are important, for they imply that *where* we choose to clean up (on high-value lands or in minority communities) and *whether* we choose to immediately redevelop undesirable land or facilities are not as important as understanding local preferences around similarity and density (recalling that the latter can include transportation efficiency).

Residential Preferences

The differences we encountered by varying the residential preference variables—similarity preferences (segregation) and density preferences—far overwhelmed the minor differences for the disamenity policy choices. Therefore, for simplicity of presenting our results, we condense the 36 scenarios to nine, varying only segregation and density preferences. We did so by averaging the outcomes of all trials across the disamenity policy choices into just the nine scenarios shown in Table 5.2.

Table 5.2 Scenario guide

	Segregation: none	Segregation: low	Segregation: high
Density: low	1	2	3
Density: mid	4	5	6
Density: high	7	8	9

Note: In reporting results for these nine scenarios, results are averaged over the minor differences caused by differences in disamenity priority and disamenity process

[10] Although we could have run each simulation longer, model-testing revealed that price and quality trends stabilized at about tick 60, so we cut off the trials at this point; if thinking in terms of years, this suggests that we began the models around roughly 1980 and concluded them at roughly 2020. See the appendix for more information.

In the graphs below we depict the time trends of the key variables, showing the lines only when the results are statistically significant (at the 95% confidence level), including aspects of the land and residential outcomes. Graphs are labeled by the scenario numbers shown in Table 5.2.

Gentrifiable and Gentrified Land We note that in all scenarios gentrifiable land was prevalent. This is not surprising because, by our definition, at each tick 25% of all land is gentrifiable. As shown in Fig. 5.2, in each scenario about 40% of land was, at some point over the course of the 60 ticks of the scenario, classified as gentrifiable, and between 20% and 35% of all land was subsequently gentrified.

There was consistently more gentrification when the residents had higher density preferences and, to a lesser extent, when they had lower similarity preferences. In low density preference scenarios (scenarios 1, 2, and 3), only about 20% of the land gentrified, while around 30% did in higher density preference settings. Land was more likely to gentrify in scenarios with moderate or high density preferences and no or low segregation preferences (Scenarios 4, 5, 7, and 8). The least amount of gentrified land for each density assumption took place in the highly segregated settings (Scenarios 3, 6, 9). Nevertheless, as Fig. 5.2 shows, in all cases, by the 60th tick between 20% and 30% of all land moved from the lowest quartile (gentrifiable) to the upper two quartiles (gentrified), showing that there was clear movement of comparatively wealthy residents into comparatively poor areas. But in no case did

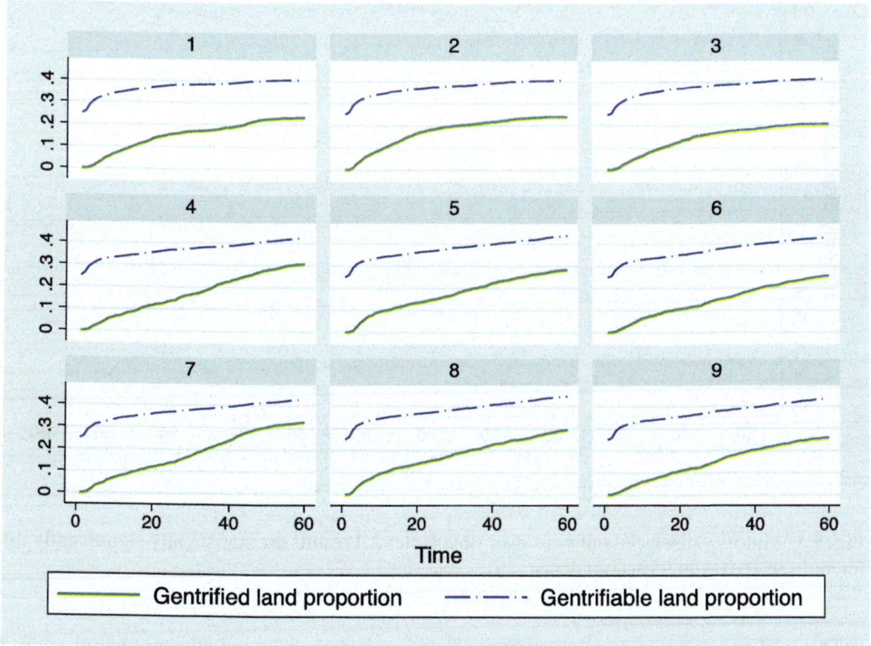

Fig. 5.2 Cumulative gentrifiable and gentrified land proportions

all the gentrifiable land gentrify. In short, Fig. 5.2 basically matches some things that we know empirically and provides some validation that our model is consistent with observations—gentrification occurs, but not all poor areas of a city will gentrify.

Displacement In Fig. 5.3, we show the proportion of residents who were displaced at least once from gentrified neighborhoods, looking first only at the displacement of minority-kind residents, because one concern around gentrification is racial displacement.[11] The minority displacement rate was highly influenced by the density preference. In low density preference settings (scenarios 1, 2, and 3), about 25–35% of minority residents were displaced at some point during the trial. In higher density preference settings, about 60–80% of minority residents were displaced at least once, with the displacement rate being *inversely* proportional to segregation preferences. About 80% of minority residents were displaced at least once in scenarios with a preference for moderate or high density but no segregation preference (scenarios 4 and 7), and only about 60% in scenarios with high levels of segregation (scenario 9, for example).

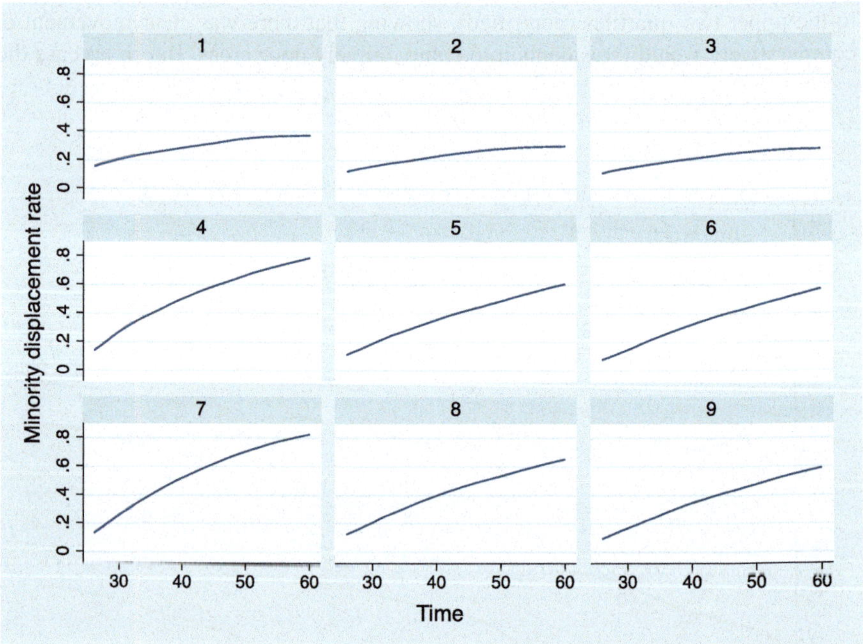

Fig. 5.3 Minority displacement rates over time Note: All results are statistically significantly different from zero with 95% confidence

[11] [This is of particular interest to us because our concern with race- and ethnicity-based environmental injustice.]

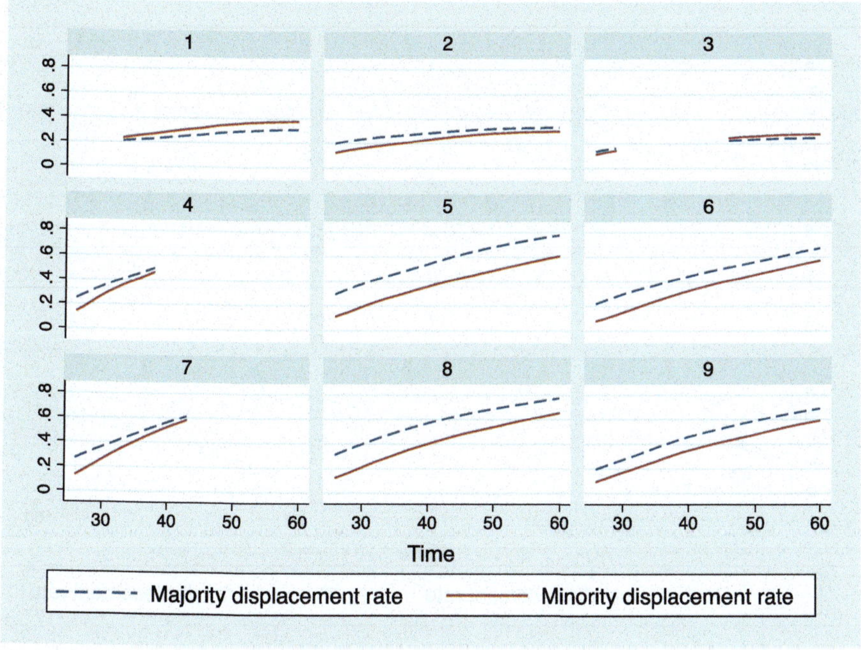

Fig. 5.4 Comparison of majority and minority displacement ratesNote: Lines are only shown when differences are statistically significant with 95% confidence

We also compare displacement rates for minority and majority residents as shown in Fig. 5.4. In Fig. 5.4, note that lines are shown only when differences between displacement rates for majorities and minorities are statistically significantly different (at the 95% confidence level). Majorities are generally displaced at higher rates than minorities in most scenarios, but the differences illustrate important distinctions. First, minority residents are only displaced at higher rates than majorities in two of the low density preference settings (1 and 3), although the rate of displacement is quite close under all of the scenarios with no segregation preference (scenarios 1, 4, and 7). Majorities are displaced at the highest rates when there is a medium or high density preference, and are displaced at higher rates than minorities when there is also a moderate to high segregation preference (scenarios 5, 6, 8, and 9).

Figure 5.4 does not differentiate residents by wealth class. If we narrow to considering only the residents in the poorest quartile of both kinds, as shown in Fig. 5.5, the conclusion is the same; we find there is little displacement of the poor when residents prefer low density (Scenarios 1, 2, and 3). Once there is a preference for density, the minority poor do worse in conditions either where there is no preference for segregation (4 and 7), or the preference for segregation is quite high, as in scenarios 6 and 9 (closer to what we observe empirically in the US), although generally speaking the majority poor do worse than the minority poor.

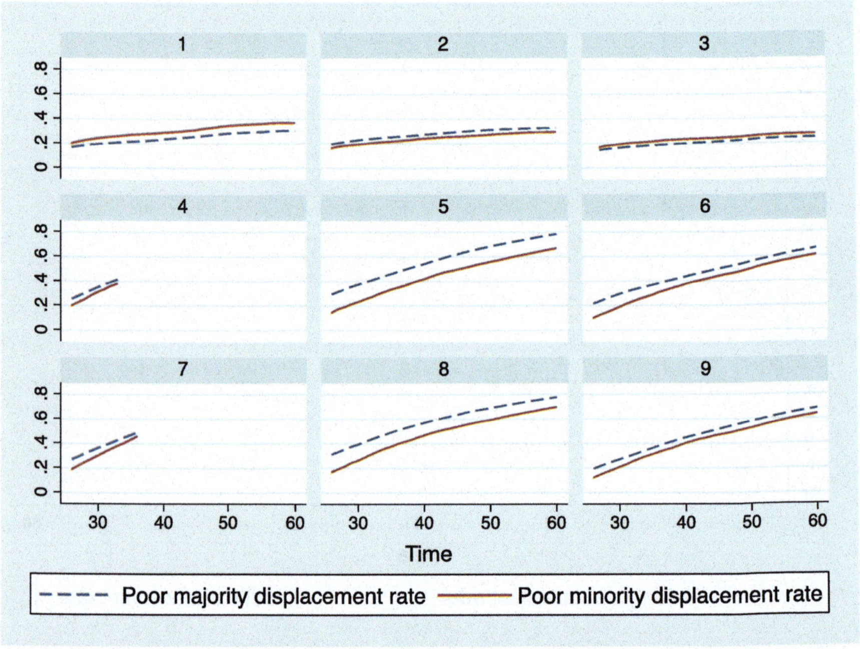

Fig. 5.5 Comparison of poor majority and poor minority displacement rates Note: Lines are only shown when differences are statistically significant with 95% confidence

The proportion of residents displaced at least once provides some insight into the burdens of gentrification-related displacement, but a fuller understanding of the burden of displacement in a dynamic housing market can be considered by the sum total number of times that poor residents were displaced, as shown in Fig. 5.6. First, poor residents are displaced very infrequently in low density preference contexts, as seen in Fig. 5.6, scenarios 1, 2, and 3 where the trend line is barely above zero. While about 35% of poor residents are displaced at least once in these scenarios (see Fig. 5.5), very few are displaced more than once. Conversely, in contexts where there is higher density, residents can be displaced multiple times, and as the city grows the number of times displaced seems to grow exponentially. In situations with low segregation preferences and higher density preferences (scenarios 4 and 7), the average number of times a poor resident is displaced during the time period of the model is six. For context, since we are assessing gentrification over 35 ticks, or the equivalent of a 35-year period, this means that poor residents in these settings were displaced about once every 6 years. Residents of both kinds are displaced more frequently when residential segregation preferences are low and less when segregation preferences are high (compare 4 with 5 and 6, and 7 with 8 and 9). Poor minority residents are displaced slightly more often than majority residents, except in the moderate segregation setting (scenarios 5 and 8).

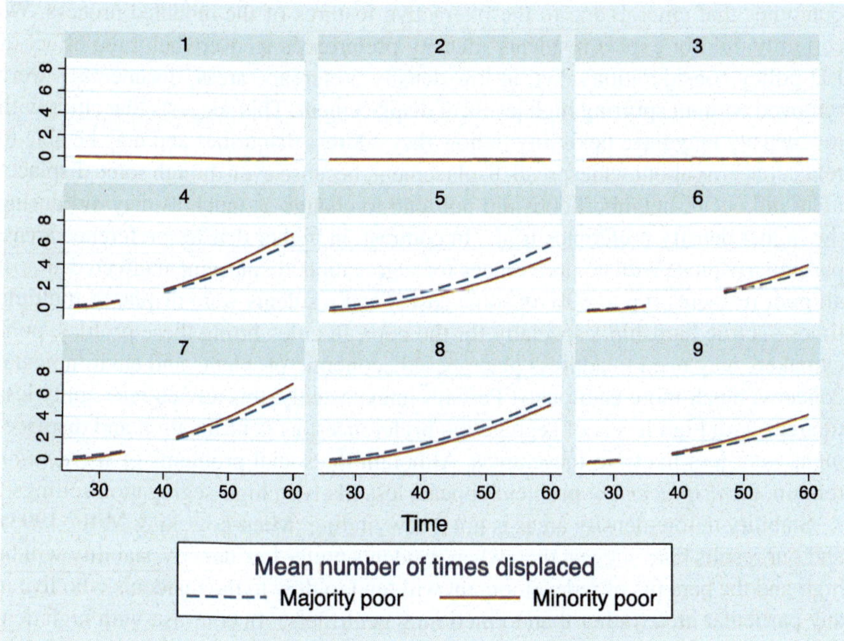

Fig. 5.6 Mean number of times poor majority and poor minority residents displaced Note: Vertical axes measures counts (rather than proportions as in Figs. 5.2 through 5.5). Lines are only shown when differences are statistically significant with 95% confidence

Discussion and Policy Implications

Although these results should be understood in the context of a stylized simulation model, our model outcomes offer some useful insight for understanding the gentrification and displacement processes and how policy may be able to address these issues. Or, more to the point, our results suggest that the same policies may result in different outcomes depending on the urban context in which the policy is being implemented, for the processes may be spurred much more by structural aspects of an urban region than by a policy response; this matches findings by Fitzgerald and Vitello (2014) when looking empirically at effects of the Community Reinvestment Act. Concerns have been raised about the potential for economic development policy to spur displacement (Checker, 2005), but this may be true only in a relatively small number of urban contexts: those with high levels of density preferences and residents who are relatively open to living in proximity to people unlike themselves. In lower density preference contexts, or where there are high levels of segregation preferences by residents, displacement seems to be a less common experience.

Our analysis suggests that appropriate policy responses to addressing economic redevelopment through environmental improvement may not be a one-size-fits-all proposition. One of the useful features of an ABM is coming up with unexpected outcomes that emerge due to the interactive features of the modeled process. We certainly did not expect residents' density preferences to overwhelm the effect of LU policy considerations. Yet, in low density preference areas, disamenities were removed without spurring high levels of displacement. This suggests that cities with low density may have flexibility in how they address disparities and may be able to relax concerns about widespread displacement, because even though some displacement did occur, neighborhoods did not tend to change as much as they did during the higher density preference trials. In contrast, in higher density preference areas, particularly those with no preferences for segregation, displacement affected almost all poor residents (up to 80% in some cases) and residents were displaced multiple times—a true hardship, especially for the poor. In cities fitting these profiles, policymakers may need to address place-based economic problems and environmental concerns much more cautiously. Perhaps unexpectedly, this advice may apply less to areas with high levels of segregation preferences, as gentrification and displacement were less likely in these areas. Although the social problems of *segregation* remain, the *displacement* problem appears less likely in high-segregation settings.

Stability in low-density areas is not a new finding (Mieszkowski & Mills, 1993), and our results here suggest that, when residents prefer low density, stability will be high and the benefits of redevelopment will tend to flow to the residents who live in any particular area (rather than to incoming gentrifiers). In contrast, with high density levels, residents are more mobile and neighborhoods are characterized by less stability. This is particularly true when residents are comfortable living around individuals different from themselves, for in these cases segregatory preferences do not provide barriers to movement. However, when residents do hold high similarity preferences (exhibit high levels of racial discrimination in housing), neighborhoods are more stable because individuals are less willing to relocate to areas with residents who are different from themselves.

Policy Implications

As noted above, our results provide less in terms of policy-specific guidelines but useful suggestions regarding the context of the policy approach. Our model suggests that environmental cleanup policy should be considered within the setting of the structural characteristics of the urban setting, with different policy recommen[d]ations for different types of cities. In Table 5.3, we propose an urban context typology that could be used to consider the probability and potential burden of gentrification-based displacement.

We consider some examples to illustrate this typology.

High Density, Low Segregation Preferences New York City may represent an area with high density and low segregation preferences (at least today, if not historically). Although the city is increasingly segregated by income class, a low segregation preference by race/ethnicity and high density preference may explain the high

Table 5.3 Gentrification context probability and burden on displacees

	Segregation preferences	
Density	High	Low
High	High probability of displacement with moderate burden on displacees	High probability of displacement with high burden on displacees
Low	Low probability of displacement with low burden on displacees	Low probability of displacement with moderate burden on displacees

rates of gentrification and displacement that have been occurring for decades there (Lees, 2003).

High Density, High Segregation Preferences Chicago may represent the high density/high segregation city, as gentrification has certainly occurred, but the well-known racial distinction between the city's north and south sides has remained stable for decades (Silver, 2015).

Low Density, High Segregation Preferences Cities with longstanding segregation, but low density, may tend to see stable neighborhoods with little gentrification or displacement, or gentrification that is very slow to take hold, as is the case in many smaller cities with stable economies including cities with good economies like Omaha, Nebraska (Gonzales, 2016), and poor economies like Stockton, California (Runner, 2016), or places in between like Indianapolis, Indiana (Taylor, 2016).

Low Density, Low Segregation Preferences Finally, low-density cities with lower (or changing) segregatory preferences, like Charleston, South Carolina (Faussett, 2015) or Durham, North Carolina, may begin to see more, albeit slowly developing, gentrification and related displacement (White, 2016).

Overall Each of these different settings may require a different policy approach to limit or compensate for displacement. Low-density cities, particularly those that are highly segregated, may not need to be very concerned with displacement as it may be less prone to occur there. This leaves them free to target removal of disamenities in minority neighborhoods.[12] Higher density cities, particularly those with comparatively low levels of segregation, may need to consider policies well beyond limitations on disamenity siting or targeted disamenity removal. While our model did not assess specific housing policies, it does suggest that in dense, diverse cities, displacement may be a significant problem.

Given the very limited effect that LU policy differences had on our outcomes (gentrification and displacement), policy in these cities may need to focus on finding a balance between providing needed economic improvements to certain

[12] Note that the finding that gentrification displacement is not affected by policy decisions regarding cleanup does not imply that other outcomes, such as environmental justice, are not affected by them.

neighborhoods, but perhaps not improvement that would appeal to a broad population, or perhaps improvements need to be balanced with other policies. For example, zoning or use restrictions, or even easements or other modes of land conservation, might be useful.

These contextualized results, which suggest that policy results depend on the context and not on a single *all else equal* outcome that can be applied to all US cities, are the sort of results that suggest the importance and value of the ABM approach, with its focus on the interactive realities of complexity. Nonetheless, our results, and these suggested policy implications, should be considered in the context of limitations of ABM as a method. First, the ABM is a highly stylized and simplified model. Its usefulness depends on the validity of the representation of agents, their interaction rules, and their environmental conditions. For any ABM, the appropriate level of abstraction and simplification is largely dependent upon the purpose of the model, and also on our current understanding of the process. Although we built the model based on our best knowledge, and tested its sensitivity to changes in parameters (see the Appendix), the assumptions are open to debate. Nevertheless, the goal of building an abstract model of gentrification and displacement is to focus on and experiment with plausible social processes, to identify the *social recipe* that leads to certain aggregate outcomes—not to estimate the effect-sizes of specific parameters as in regression analysis. In fact, since the model relaxes the typical statistical assumption of the independence of input variables, standard effect size considerations are not applicable. The model assumes the interdependence of all input variables and thus any particular scenario result is an outcome of an interaction between variables. Any assessment of the *independent* effects of these interdependent variables would provide a false picture of the nature of the model. Though stylized, the formal dynamism of the ABM leads to understanding of social processes that can lead to certain outcomes, allowing us to craft more context-specific policies that take into account the complex dynamics of the social world and less content-specific policies that assume the indepen[den]ce of variables that, we argue, are not independent.

Although our model suggests that gentrification and policies to address displacement disparities probably ought to be considered within the context of segregation and density trends of a city, we do not suggest that these considerations should be the only considerations. Nevertheless, being able to describe our results in reference to real cities illustrates the potential utility of the model's insights.

A final limitation is our operationalization of racial distinctions. In our model, difference is a relatively benign concept. It is a distinction of mathematical majority/minority with some discriminatory preference expressed through the residential similarity preference. With simple agents that possess no social memory and simplistic behavioral assumptions, we do not model anything as complex as actual racial disparity and segregation—the agents have no institutional memory of historical racism or contemporary discrimination—though the accretion of these realities may be instantiated in a very simple way via the similarity preference.

Conclusion

In this research, we model gentrification and gentrification-related displacement of residents. We do so through the development of an ABM of a simple urban region, considering different urban contexts and policy approaches to improving environmental conditions and the relationship of these policies to subsequent gentrification and displacement. We find that gentrification-related displacement is most likely, and most potentially burdensome, in urban regions characterized by high levels of density and low levels of residential segregation preferences. Displacement is far less prevalent in low-density regions, particularly those with high segregation preferences. We find that choosing to clean up disamenities based on land value or minority area makes little difference to these outcomes, as does whether the disamenity is simply removed (and the land is thus available in the market) or is immediately replaced with an amenity. Our model indicates that the characteristics of the region may affect the policy outcome more than any particular policy approach. In areas with high displacement, any policy to improve environmental conditions fostered environmental improvement that disproportionately benefited incomers, while in low-displacement regions any policy tended to effectively address disparities in living conditions. Though disamanity cleanup may be well-intentioned, its impacts for residents appears to depend on the local context, and decision-makers should take this into account before acting.

Appendix

Appendix 5A Model robustness assessments

Parameters	Model value	Range of values tested	Result
Land uses			
Number of amenities/disamenities per resident	10, fixed	10–50 by intervals of 10	At higher numbers, fewer firms were established, thus less quality variability and generally less gentrification. Ten provides a good balance between a reasonable assumption and enough variation in quality and resident proximity to land uses. Results at high levels showed few discernable trends, but as the number of residents per land use was lowered, similar patterns to those presented emerged

(continued)

Appendix 5A (continued)

Parameters	Model value	Range of values tested	Result
Residential proximity preference	0–10, uniform distribution	0–10, from normal distribution ~ (5, 2)	Normal distribution resulted in many firms near 5, and little variation in residential proximity and its effects. Results were similar but muted
Location preference/ cleanup priority	Lowest price; near minority residents	Random, low price, near minority residents, away from majority residents	Since the minority population was assumed to be less wealthy than the majority, model results were robust to any location preference or cleanup priority for land uses
Residents			
Similarity preference	0, 0.20, 0.50/0.80 as a location constraint	0.20, 0.50, 0.80; similarity preference part of utility calculation rather than constraint.	Results with similarity preference in the utility function were conceptually the same as those presented, with a dampened effect. It was left as a constraint to remain consistent with Schelling (1971)
Utility preferences	Normal distribution assignment; equal weighting in utility function; price, quality, density, and transportation access	Combinations of different weighting; removal and addition of some preferences; including density as a constraint rather than a preference	Different weighting of the preferences had little effect on results. Density as a constraint enhanced the effect of density, but literature suggested a preference in the utility function as more justified
Patches			
Transportation	Proximity to roads as a resident preference	Proximity to roads not a resident preference	A preference to be located near transportation routes muted the effects of land-use changes on gentrification, but enhanced it near transportation routes. The preference was included as it is realistic to assume that areas with transportation access would be prone to gentrification
Price	Random draw from gamma distribution of US Census values	Random draw from normal distribution	The distribution type made little difference in model results. Gamma distribution used as the more justifiable option

(continued)

Appendix 5A (continued)

Parameters	Model value	Range of values tested	Result
Price and quality diffusion	Based on proximity to LUs and wealth of nearby populations	Based on quality; diffusion of prices and quality from nearby plots	Reliance of quality as a price mechanism quickly made quality patches cost too much for even wealthy residents. Proximity to LUs and wealth led to somewhat stable, but not completely stable neighborhoods, which was determined to be realistically valid
Global			
Ticks	60	Range: 50–200 by intervals of 25	Although somewhat variable between about 50 and 100, the trends presented tended to continue on similar trajectories after tick 60
Period before policies implemented	25 ticks	Range: 10–70 ticks, by intervals of 10	Environmental justice problems only became more significant if more ticks passed before policy implementation. Tick 25 seemed a good choice, as trends were clear but not yet overwhelming
World size	40 × 40	10 × 10 to 100 × 100 in intervals of 10	Trends were similar regardless of world size. Emergence of gentrification was faster in smaller worlds and slower in larger worlds. A 40 × 40 world was a nice balance between a sizable city that did not overtax available computing resources
Population growth rate	5%	0–25%	Results were similar regardless of population growth rate, although at higher growth rates space became more constrained, so initially the trends were amplified both in the setup and policy implementation phases, with trends leveling out earlier than 60 ticks

References

Bartik, T. J. (1991). *Who benefits from state and local economic development policies?* W.E. Upjohn Institute for Employment Research. https://doi.org/10.17848/9780585223940

Batty, M. (2007). *Cities and complexity: Understanding cities with cellular automata, agent-based models and fractals*. MIT Press.

Beauregard, R. (1986). The chaos and complexity of gentrification. In N. Smith & P. Williams (Eds.), *The gentrification of the city* (pp. 35–55). Allen and Unwin.

Black, J. (1980). Private market housing renovation in central cities: An urban land institute survey. In S. Laska & S. Spain (Eds.), *Back to the city: Issues in neighborhood renovation* (pp. 3–12). Pergamon.

Brown, D., & Robinson, D. (2006). Effects of heterogeneity in residential preferences on an agent-based model of urban sprawl. *Ecology and Society, 11*(1), 46. http://www.jstor.org/stable/26267798

Brueckner, J., & Rosenthal, S. (2009). Gentrification and neighborhood housing cycles: Will America's future downtowns be rich? *Review of Economics and Statistics, 91*(4), 725–743. https://doi.org/10.1162/rest.91.4.725

Campbell, H., Kim, Y., & Eckerd, A. (2015). *Rethinking environmental justice in sustainable cities: Insights from agent-based modeling*. Routledge.

Checker, M. (2005). *Polluted promises: Environmental racism and the search for justice in a southern town*. New York University Press.

Checker, M. (2011). Wiped out by the "greenwave": Environmental gentrification and the paradoxical politics of urban sustainability. *City & Society, 23*(2), 210–229. https://doi.org/10.1111/j.1548-744X.2011.01063.x

Chen, X., & Zhan, F. (2008). Agent-based modeling and simulation of urban evacuation: Relative effectiveness of simultaneous and staged evacuation strategies. *Journal of the Operational Research Society, 59*(1), 25–33. https://doi.org/10.1057/palgrave.jors.2602321

Clark, W. A. V. (1992). Residential preferences and residential choices in a multiethnic context. *Demography, 29*(3), 451–466. https://doi.org/10.2307/2061828

Clark, E. (2005). The order and simplicity of gentrification: A political challenge. In R. Atkinson & G. Bridge (Eds.), *Gentrification in a global context: The new urban colonialism*. Routledge.

Clay, P. (1979). *Neighborhood renewal: Middle class resettlement and incumbent upgrading in American neighborhoods*. DC Heath.

Dean, J.S., Gumerman, G.J., Epstein, J.M., Axtell, R.L., Swedlund, A.C., Parker, M.T., & McCarroll, S. (2000). Understanding Anasazi culture change through agent-based modeling. In G.J. Gumerman and T. A. Kohler (Eds.), *Dynamics in human and primate societies: Agent -based modeling of social and spatial processes* (pp.179–206). Oxford University Press.

Devisch, O., Timmermans, J., Arentze, T., & Borger, A. (2009). An agent-based model of residential choice dynamics in nonstationary housing markets. *Environment and Planning A, 41*(8), 1997–2013. https://doi.org/10.1068/a41158

Dieleman, F., & Wegener, M. (2004). Compact city and urban sprawl. *Built Environment, 30*(4), 308–323. https://doi.org/10.2148/benv.30.4.308.57151

Eckerd, A. (2011). Cleaning up without clearing out? A spatial assessment of environmental gentrification. *Urban Affairs Review, 47*(1), 31–59. https://doi.org/10.1177/1078087410379712

Eckerd, A., & Reames, T. (2012). Urban renaissance or invasion: Planning the development of a simulation model of gentrification. In A. Desai (Ed.), *Simulation for policy inquiry* (pp. 83–100). Springer.

Engelen, G., White, R., Uljee, I., & Drazan, P. (1995). Using cellular automata for integrated modeling of socio-economic environments. *Environmental Monitoring and Assessment, 34*(2), 203–214. https://doi.org/10.1007/BF00546036

Epstein, J. M. (2006). *Generative social science: Studies in agent-based computational modeling*. Princeton University Press. https://doi.org/10.1515/9781400842872

Fausset, R. (2015). Steeped in Racial History, Charleston Ponders Its Future. The New York Times. https://www.nytimes.com/2015/06/20/us/steeped-in-racial-history-charleston-reflects-on-its-relationship-with-blacks.html

Filatova, T., Parker, D., & Van der Veen, A. (2009). Agent-based urban land markets: Agent's pricing behavior, land prices and urban land use change. *Journal of Artificial Societies and Social Simulation, 12*(1), 3.

Fitzgerald, J., & Vitello, S. P. (2014). Impacts of the community reinvestment act on neighborhood change and gentrification. *Housing Policy Debate, 24*(2), 446–466. https://doi.org/10.1080/10511482.2013.858364

Freeman, L. (2008). Comment on 'the eviction of critical perspectives from gentrification research'. *International Journal of Urban and Regional Research, 32*(1), 186–191. https://doi.org/10.1111/j.1468-2427.2008.00771.x

Freeman, L. (2011). *There goes the 'hood: Views of gentrification from the ground up*. Temple University Press.

Fusch, R. (1980). A case of too many actors? Columbus. In S. Laska & S. Spain (Eds.), *Back to the city: Issues in neighborhood renovation* (pp. 156–172). Pergamon.

Glass, R. (1964). London. In Center for Urban Studies (Ed.), *London: Aspects of change*. MacGibbon and Kee.

Gonzalez, C. (2016, April 11). Guiding gentrification: Nonprofit worries about downside to rising rents for those less well-off. *Omaha World Herald*. https://omaha.com/money/guiding-gentrification-nonprofit-worries-about-downside-to-rising-rents-for-those-less-well-off/article_795551bd-99f7-5c80-b7bf-1ebc4c28eecc.html

Grossmann, K., & Haase, A. (2016). Neighborhood change beyond clear storylines: What can assemblage and complexity theories contribute to understandings of seemingly paradoxical neighborhood development? *Urban Geography, 37*(5), 727–747. https://doi.org/10.1080/02723638.2015.1113807

Haase, D., Lautenbach, S., & Seppelt, R. (2010). Modeling and simulating residential mobility in a shrinking city using an agent-based approach. *Environmental Modeling & Software, 25*(10), 1225–1240. https://doi.org/10.1016/j.envsoft.2010.04.009

Hammel, D., & Wyly, E. (1996). A model for identifying gentrified areas with census data. *Urban Geography, 17*(3), 248–268. https://doi.org/10.2747/0272-3638.17.3.248

Hamnett, C. (1991). The blind men and the elephant: The explanation of gentrification. *Transactions of the Institute of British Geographers, 16*(2), 173–189. https://doi.org/10.2307/622612

Helms, A. (2003). Understanding gentrification: An empirical analysis of the determinants of urban housing renovation. *Journal of Urban Economics, 54*(3), 474–498. https://doi.org/10.1016/S0094-1190(03)00081-0

Holland, J. H. (1998). *Emergence*. Perseus Books.

Huang, Q., Parker, D., Filatova, T., & Sun, S. (2014). A review of urban residential choice models using agent-based modeling. *Environment and Planning B, 41*(4), 661–689. https://doi.org/10.1068/b120043p

Hwang, J. (2016). While some things change, some things stay the same: Reflections on the study of gentrification. *City & Community, 15*(3), 226–230. https://doi.org/10.1111/cico.12188

Hwang, J., & Sampson, R. (2014). Divergent pathways of gentrification: Racial inequality and the social order of renewal in Chicago neighborhoods. *American Sociological Review, 79*(4), 726–751. https://doi.org/10.1177/0003122414535774

Jackson, J., Forest, B., & Sengupta, R. (2008). Agent-based simulation of urban residential dynamics and land rent change in a gentrifying area of Boston. *Transactions in GIS, 12*(4), 475–491. https://doi.org/10.1111/j.1467-9671.2008.01109.x

Kennedy, M., & Leonard, P. (2001). *Dealing with neighborhood change: A primer on gentrification and policy choices*. Discussion paper prepared for The Brookings Institution Center on Urban and Metropolitan Policy.

Kirschenbaum, A. (1996). Residential ambiguity and relocation decisions: Population and areas at risk. *International Journal of Mass Emergencies and Disasters, 14*(1), 79–96. https://doi.org/10.1177/028072709601400105

Kolko, J. (2007). *The determinants of gentrification*. SSRN. https://ssrn.com/abstract=985714. https://doi.org/10.2139/ssrn.985714.

Lees, L. (2003). Super-gentrification: The case of Brooklyn Heights, new York City. *Urban Studies, 40*(12), 2487–2509. https://doi.org/10.1080/0042098032000136174

LeGates, R., & Hartman, C. (1986). The anatomy of displacement in the U.S. In N. Smith & P. Williams (Eds.), *The gentrification of the city* (pp. 178–200). Allen and Unwin.

Ley, D. (1981). Inner city revitalization in Canada: A Vancouver case study. *Canadian Geographer, 25*(2), 124–148. https://doi.org/10.1111/cag.1981.25.2.124

Ley, D. (1986). Alternative explanations for inner-city gentrification: A Canadian assessment. *Annals of the Association of American Geographers, 76*(4), 521–535. https://doi.org/10.1111/j.1467-8306.1986.tb00134.x

Lyons, M. (1996). Gentrification, socioeconomic change, and the geography of displacement. *Journal of Urban Affairs, 18*(1), 39–62. https://doi.org/10.1111/j.1467-9906.1996.tb00365.x

Mieszkowski, P., & Mills, E. S. (1993). The causes of metropolitan suburbanization. *Journal of Economic Perspectives, 7*(3), 135–147. https://doi.org/10.1257/jep.7.3.135

Miller, J. H., & Page, S. E. (2007). *Complex adaptive systems: An introduction to computational models of social life*. Princeton University Press.

Nagel, K., & Rickert, M. (2001). Parallel implementation of the TRANSIMS micro-simulation. *Parallel Computing, 27*(12), 1611–1639. https://doi.org/10.1016/S0167-8191(01)00106-5

Newman, K., & Wyly, E.K. (2006, July). *Gentrification and displacement revisited: A fresh look at the New York City experience*. Centre for Urban and Community Studies Research Bulletin #31. University of Toronto. http://neighbourhoodchange.ca/wp-content/uploads/2011/05/CUCSRB31-NewmanWyly.pdf

O'Sullivan, D. (2002). Toward micro-scale spatial modeling of gentrification. *Journal of Geographic Systems, 4*(3), 251–274. https://doi.org/10.1007/s101090200086

O'Sullivan, D. (2009). Changing neighborhoods—Neighborhoods changing. *Sociological Methods & Research, 27*(4), 498–530. https://doi.org/10.1177/0049124109334793

Rayle, L. (2015). Investigating the connection between transit-oriented development and displacement: Four hypotheses. *Housing Policy Debate., 25*(3), 531–548. https://doi.org/10.1080/10511482.2014.951674

Rose, D. (1984). Rethinking gentrification: Beyond the uneven development of Marxist urban theory. *Society and Space, 2*(1), 47–74. https://doi.org/10.1068/d020047

Runner, M. (2016). Don't count Stockton out. *Comstock's*. https://www.comstocksmag.com/commentary/dont-count-stockton-out

Sanchez, T., & Dawkins, C. (2001). Distinguishing city and suburban movers: Evidence from the American housing survey. *Housing Policy Debate, 12*(3), 607–631. https://doi.org/10.1080/10511482.2001.9521420

Schaffer, R., & Smith, N. (1986). The gentrification of Harlem? *Annals of the Association of American Geographers., 76*(3), 347–365. https://doi.org/10.1111/j.1467-8306.1986.tb00124.x

Schelling, T. C. (1971). *Micromotives and macrobehavior*. W.W. Norton.

Schill, M., & Nathan, R. (1983). *Revitalizing America's cities: Neighborhood reinvestment and displacement*. State University of New York Press.

Silver, N. (2015). The most diverse cities are often the most segregated. *FiveThirtyEight*. https://fivethirtyeight.com/features/the-most-diverse-cities-are-often-the-mostsegregated/

Slater, T. (2006). The eviction of critical persepectives from gentrification research. *International Journal of Urban and Regional Research, 30*(4), 737–757. https://doi.org/10.1111/j.1468-2427.2006.00689.x

Smith, N. (1979). Toward a theory of gentrification: A back to the city movement by capital, not people. *Journal of the American Planning Association, 45*(October), 538–548. https://doi.org/10.1080/01944367908977002

Sullivan, D. M. (2006). Assessing residents' opinions on changes in a gentrifying neighborhood: A case study of the Alberta neighborhood in Portland, Oregon. *Housing Policy Debate, 17*(3), 595–624. https://doi.org/10.1080/10511482.2006.9521583

Sullivan, D. M. (2007). Reassessing gentrification: Measuring residents' opinions using survey data. *Urban Affairs Review, 42*(4), 583–592. https://doi.org/10.1177/1078087406295828

Taylor, E. (2016, Decmber 14). Are Indy neighborhoods gentrifying? *Nuvo*. https://www.nuvo.net/arts/general_arts/are-indy-neighborhoods-gentrifying/article_f7ef3602-5992-5043-b61a-25a88b59aabd.html

Torrens, P., & Nara, A. (2007). Modeling gentrification dynamics: A hybrid approach. *Computers, Environment and Urban Systems, 31*(3), 337–361. https://doi.org/10.1016/j.compenvurbsys.2006.07.004

Valler, D., & Wood, A. (2010). Conceptualizing local and regional economic development in the USA. *Regional Studies, 44*(2), 139–151. https://doi.org/10.1080/00343400802378768

Veldkamp, A., & Verburg, P. H. (2004). Modeling land use change and environmental impact. *Journal of Environmental Management, 72*(1–2), 1–3. https://doi.org/10.1016/j.jenvman.2004.04.004

Vigdor, J. (2002). Does gentrification harm the poor? *Brookings-Wharton Papers on Urban Affairs*, 133–174. https://www.jstor.org/stable/25067387

White, G.B. (2016). The downside of Durham's rebirth. *The Atlantic*. https://www.theatlantic.com/business/archive/2016/03/the-downside-of-durhamsrebirth/476277/

Wyly, E., & Hammel, D. (1998). Modeling the context and contingency of gentrification. *Journal of Urban Affairs, 20*(3), 303–326. https://doi.org/10.1111/j.1467-9906.1998.tb00424.x

Chapter 6
Clusters of Conditions for Green Gentrification

Jieun Kim and Michelle Stuhlmacher

Abstract Green gentrification suggests that new urban greening projects in deprived neighborhoods can encourage gentrification. Does it mean that deprived neighborhoods without new greening are not or are less prone to gentrification? To examine the role of new greening in deprived neighborhoods, we rely on a novel set-theoretic method, Qualitative Comparative Analysis. This method is suitable to ask and answer under what conditions green gentrification occurs. We analyzed 26 Chicago neighborhoods and compared ones with a new (large or small) park and ones without a new park and their gentrification status between 2010 and 2020. We find that (1) a large new park alone is *not necessary to encourage* gentrification. (2) Depending on neighborhood conditions (e.g., lack of pre-existing greenness), a *small* new park can encourage gentrification. (3) When a neighborhood has nearby areas that were gentrified in the previous decade, the neighborhood *without a new park* was gentrified even with relatively more attention of city government on affordable housing in the community area. We recommend city officials and planners account for neighborhood contexts when considering greenspace provisions to achieve just planning outcomes.

Introduction

The value of urban greening in bringing diverse benefits to health, environment, and climate resilience has been well documented (Bai et al., 2013; Kabisch et al., 2016; McCormack et al., 2010; Sugiyama et al., 2013). Especially as a means of improving the environmental conditions of highly built urban cores, urban greening can be

J. Kim (✉)
School of Public Affairs, Arizona State University, Phoenix, AZ, USA
e-mail: jkim593@asu.edu

M. Stuhlmacher
Department of Geography & GIS, DePaul University, Chicago, IL, USA
e-mail: michelle.stuhlmacher@depaul.edu

H. E. Campbell et al., *Green Gentrification and Environmental Injustice*,
https://doi.org/10.1007/978-3-031-65100-7_6

initiated for the benefit of low-income or working minorities who live in these largely impoverished neighborhoods (Gould & Lewis, 2017). However, the paradox of urban greening is that intended beneficiaries (i.e., existing low-income residents) may become displaced by new middle- and upper-class residents as investments in urban greening increase property and rent values of the surrounding neighborhoods attracting real-estate investments (Checker, 2011). This phenomenon, known as "green gentrification," is complex to conceptualize and challenging to measure, specifically when it comes to quantifying the neighborhood change and characteristics of urban greening, such as types, vegetation, and aesthetics (Easton et al., 2020; Quinton et al., 2022; Zuk et al., 2018).

Recently, scholars began to tap into the complexity of green gentrification by teasing out the temporal order of the causal relationship between urban greening and gentrification (Rigolon & Collins, 2022), different roles of urban greening in the gentrification process (Anguelovski et al., 2022), and joint causes of urban greening and neighborhood contexts that could encourage or deter gentrification (J. Kim et al., 2024). Deepening our understanding of green gentrification is a crucial task because urban greening is a key part of the climate adaptation strategies that cities began employing to alleviate climate change impacts (Bona et al., 2023). Further, cleaning and greening has valuable health benefits. As cities around the world are increasingly investing in urban greenspace (Mabon & Shih, 2021; Mell et al., 2013), a more nuanced understanding of conditional causes that encourage green gentrification would help city officials and planners to navigate ways to balance on the tightrope of achieving just greening in the face of climate change depending on their local contexts. This study, therefore, examines under what conditions new urban greening projects encourage gentrification. It focuses on analyzing urban greening decisions (i.e., adding a new park) in Chicago's downtown neighborhoods between 2010 and 2020.

Green Gentrification Research

From the literature, we identify four key factors that contribute to green gentrification. They include the addition of a new park in an economically distressed neighborhood, which is a central factor of green gentrification discussions, and three other crucial factors: pre-existing greenspace before a new park is added; affordable housing units, especially assisted by governments; and economic spillover from nearby neighborhoods.

New Park The key argument of green gentrification centers around whether a new greening project in a deprived neighborhood is a main driver of gentrification. The point of interest is the newly inaugurated greenspace. While greenspace differs in physical size, type, and other characteristics, the most examined characteristic is the size of a new greenspace. A handful of studies examine large-scale urban greening projects in the U.S. and other countries and found that large parks are linked with gentrification (Anguelovski et al., 2018; Chen et al., 2021; Immergluck & Balan, 2018; Kwon et al., 2017; Loughran, 2014; Smith et al., 2016). In an effort to find ways to achieve anti-displacement outcomes of greening, some scholars posited that smaller parks may mitigate the gentrification and displacement effect of green investment (Wolch et al., 2014). However, Rigolon and Németh (2020) found that small new parks in proximity to city centers encourage gentrification. Their finding suggests that new parks, regardless of size, might encourage gentrification *contingent upon* other conditional factors, such as neighborhood characteristics.

Even before we discuss whether the size of a new park encourages gentrification, there is a fundamental question that we need to attend to: Is new greening a difference-maker for gentrification? The question is still salient in the green gentrification literature because greening, especially in low-income neighborhoods, brings not only environmental and health benefits but also addresses the environmental inequality gap. If "greening alone is the main intervention in the built environment that explains gentrification" (Anguelovski et al., 2022, p. 5), we may see growing fear of green gentrification among low-income residents, preventing them from receiving the diverse benefits of greening.[1] This urges us to sharpen our understanding about the greening-gentrification relationship by examining whether neighborhoods without a new park, but similar in characteristics to neighborhoods with a new park, undergo the gentrification process. If we observe gentrification with and without new greening, we need to explore subtler explanations about how neighborhood conditions and different greening scenarios are combined to explain gentrification.

Pre-existing Greenspace Anguelovski et al. (2022) tested whether greening plays different roles in explaining gentrification: (1) leading role: playing the standout explanatory role of gentrification, (2) integrated role: having an explanatory role that is roughly on par with that of new developments, and (3) subsidiary role: playing a secondary role along with other built environment improvements for gentrification. They found that "greening, and greening alone," played a central role in encouraging gentrification in 8 out of 17 studied cities that experienced gentrification, such as Atlanta, Nantes, and Vancouver (Anguelovski et al., 2022). Interestingly, these cities commonly had large new parks, which confirms findings of previous literature regarding a large new park having a statistically meaningful relationship with gentrification. However, these study results then lead to the question of greenspace inequality: Should disadvantaged communities entirely avoid having a large new park in fear of gentrification? It merely leaves us with anti-gentrification options suggested by just green enough (JGE) scholars: providing small parks, or other community-driven greening initiatives, which may be less attractive to real-estate developers and other outside investers (Curran & Hamilton, 2012). From the greenspace-equity point of view, it becomes exigent to ask if the gentrification

[1] As mentioned in earlier chapters, this has been known to occur in at least some cases.

outcome of large new parks differs depending on the neighborhood's existing conditions, especially the initial greenspace coverage. The overall greenness of the neighborhood –nature-rich or nature-poor– is suggested to play a role in the gentrification process (Anguelovski et al., 2019; Sharifi et al., 2021; Stuhlmacher et al., 2022). While the effect of new green investment on gentrification in the context of differing degrees of greenness has not yet been empirically examined, it may be a critical contextual factor that differentiates the gentrification outcome of a large new park.

Government-Assisted Housing Units An important neighborhood condition specific to green gentrification is the provision of affordable housing units to curb the displacement of existing low-income residents (Curran & Hamilton, 2012; Rigolon & Németh, 2018; Wolch et al., 2014; Curran et al., 2023). JGE and other green gentrification scholars posit that affordable housing programs will help existing low-income residents remain in the neighborhoods that undergo gentrification due to greening (Curran & Hamilton, 2012; Gould & Lewis, 2017). However, empirical evidence of affordable housing having a deterring role in the gentrification process suggests otherwise. For example, Rigolon and Németh (2020) found a positive statistical relationship between the percentage of federally subsidized housing units and gentrification. Without a deeper understanding of the history of the affordable housing program, this result may be baffling. Federal housing subsidy programs like HOPE VI were designed to deconcentrate poverty through the relocation of low-income residents into mixed-income neighborhoods and redeveloping the impoverished neighborhoods (Oakley & Burchfield, 2009). Such efforts intentionally displace low-income residents and encourage gentrification. While some affordable housing programs may not play a deterring role in gentrification as envisioned, other programs provided with the government's specific intention to curb housing unaffordability during the urban redevelopment process may help existing low-income residents to stay. Therefore, the effect of affordable housing programs on the gentrification process is highly dependent upon the intention and implementation of government programs and needs careful exploration to examine how it interacts with the development of the surrounding built environment.

Economic Spillover Economic growth and new residential development were found to have positive roles in explaining gentrification in studies that examined 28 cities globally (Anguelovski et al., 2022). Other studies looking at neighborhoods in US cities, including Chicago, found that economic growth and socio-demographic changes in neighborhoods influence the gentrification process of their adjacent neighborhoods (Brazil & Portier, 2022; Guerrieri et al., 2013). This economic spillover effect is a manifestation of the desire to live close to a more affluent neighborhood that increases the housing prices in bordering neighborhoods (Guerrieri et al., 2013). Additionally, the economic spillover effect indicates that neighborhoods of interest may be under the influence of large-scale urban redevelopment efforts. The influence of the large-scale redevelopment efforts may have been manifested through the existence of nearby neighborhoods whose gentrification process has

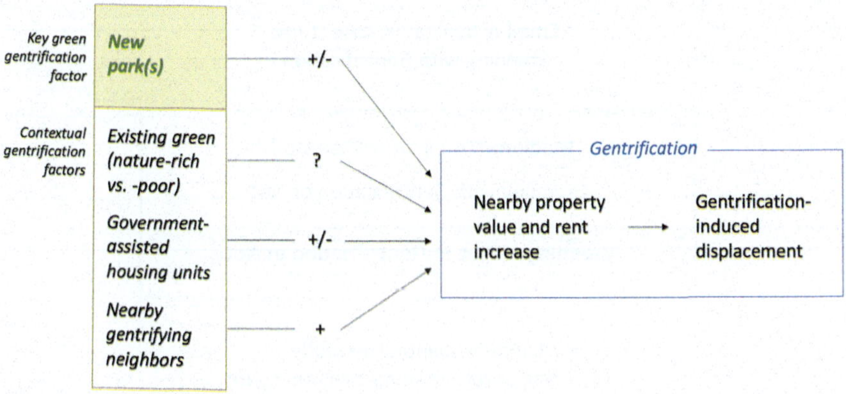

Fig. 6.1 Research framework. (*Source*: The authors)

already begun in the previous decade, and that may interfere with understanding the effect of new greening initiatives for encouraging gentrification in neighborhoods.

Figure 6.1 summarizes the literature review. Symbols in the middle of the figure show our literature review focusing on empirical evidence. For example, large parks can be a factor that encourages or promotes (indicated by a plus sign; +) gentrification whereas small parks can be a factor that discourages or deters (indicated by a minus sign; −) gentrification. Depending upon the goal of the government's housing assistance programs, it can also encourage or discourage gentrification displacement. The role of pre-existing greenspace was recently theoretically suggested but is still lacking empirical evidence (indicated by a question mark; ?). The literature suggests that economic spillover encourages gentrification. To our knowledge, there is no research that shows economic spillover discourages gentrification. We use this research framework to analyze the case of Chicago neighborhood green gentrification. By doing so, we can have a more contextualized understanding about how these factors are intertwined, specific to Chicago, and in similar cities.

Qualitative Comparative Analysis (QCA) Approach to Green Gentrification

Green gentrification has been debated upon empirical grounds. Qualitative case studies focused on a single or a few parks argue that new parks, especially large parks, encourage gentrification (Derickson et al., 2021; Zuk et al., 2018). These studies of green gentrification tend to focus on one or a few cases with intense examinations to understand the phenomenon in detail (e.g., Bryson, 2012; Checker, 2011; Curran & Hamilton, 2012; Pearsall, 2010). The in-depth examination of the qualitative research is time-consuming and thus does not allow researchers to compare cases that are similar in nature but did not experience gentrification with or

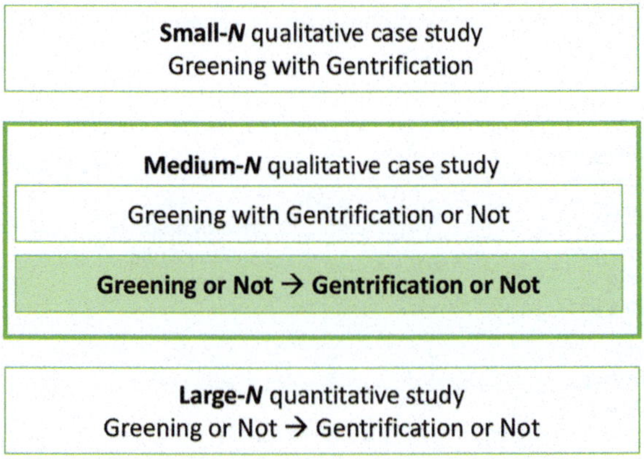

Fig. 6.2 Research approaches to green gentrification. (*Source:* The authors)

without new greening. Hence, the research design for the case selection is limited to ex-post-facto cases with new greening that surely experienced gentrification.

Unlike qualitative studies focusing on true positive cases (greening with gentrification in Fig. 6.2), quantitative studies examine green gentrification through statistical analyses that include both positive and negative cases. However, in quantitative studies, a large number of negative cases shrouds the effect of a few positive cases, especially when studies conduct citywide analyses (e.g., Stuhlmacher et al., 2022; Anguelovski et al., 2022). These studies show mixed evidence on the statistical association between new greening and gentrification (e.g., Rigolon & Németh, 2020; Sharifi et al., 2021; Stuhlmacher et al., 2022; Triguero-Mas et al., 2022). These limitations of both qualitative and quantitative approaches require a middle ground that utilizes medium-N cases based on qualitative knowledge and a comparative technique for analysis.

We use a different approach to compare cases, one that is especially suitable for examining the *combined* effect of urban greening and gentrification-related variables in triggering gentrification. Pertinent to such a purpose, Qualitative Comparative Analysis (QCA) finds the constellations of causal conditions (i.e., configurations) that are sufficient to bring about the social phenomenon by systematically comparing cases and examining set relations (Schneider & Wagemann, 2012). As mentioned in Chap. 3, QCA embraces causal complexity embedded within social realities. The emergence of a social phenomenon such as gentrification is not necessarily the product of a single condition (e.g., greening) but that of interactions among several conditions that are meaningful to the context. However, the fact that a social phenomenon was the consequence of causal mechanisms consisting of multiple interacting variables specific to context does not necessarily mean that those mechanisms are always observable. The term 'causal' here indicates not a universal law but specific *temporal* and *local* mechanisms, producing a specific social reality (Verweij, 2015).

By strengthening the research design of qualitative case studies, QCA as a research approach to green gentrification can bridge the research gap existing between small-N qualitative case studies (i.e., greening with gentrification cases only) and large-N quantitative studies that examine whether greening is associated with or causes gentrification. Figure 6.2 illustrates two possible scenarios of a medium-N case study that can improve our knowledge of green gentrification. The first medium-N research design possibility is similar to small-N case studies but with an increased number of cases and a systematic comparison using set theory and Boolean algebra (i.e., QCA). In our previous work, for example, we performed a medium-N case study using this framework and analyzed only the cases of new greening in Chicago neighborhoods, in order to understand the role of park size in the *presence and absence* of gentrification (J. Kim et al., 2024). Still, this research design does not answer the counterfactual question "What if no park?". We do not know whether new greening was the difference-maker for gentrification to occur. The second medium-N research design in Fig. 6.2 improves the previous QCA research design by adding cases *without new greening* as a comparison group. This improvement in the QCA research design upon which we build our analysis in this chapter allows us to investigate the role of greening on gentrification as a difference-maker, using three different choices of new park cases: large new park, small new park, and no new park.

QCA as a Method

In QCA, set relations are interpreted in terms of *necessity* and *sufficiency*. A condition (e.g., a new park) is *necessary* for the outcome (e.g., gentrification) if the condition is always present when the outcome occurs and if the outcome is not present without the condition. On the other hand, a condition is *sufficient* for the outcome, if the outcome is present whenever the condition is present; the outcome can still be present even if the condition is not present. QCA uses logical operators to analyze set relations and does not to yield the probabilities, average effects, or magnitudes of conditions as in statistical analysis.

If a single condition is present for the outcome to occur, it is necessary and sufficient to cause the outcome by itself. Most of the time, social reality does not occur due to a single condition but in the context of other conditions. Therefore, a condition is insufficient itself to explain the outcome condition but a necessary part of a combination of conditions that is itself unnecessary but sufficient for the outcome condition: Insufficient but Necessary part of an Unnecessary but Sufficient condition (INUS) (Goertz, 2003; Schneider & Wagemann, 2012). INUS allows "different causally relevant conditions [to be] combined in a variety of ways to produce a given outcome" (Verweij, 2015, p. 49).

Procedurally, QCA first assigns each case a membership status for a condition by drawing a boundary that defines zones of inclusion and exclusion as well as a cutoff point to be a member of the condition (Schneider & Wagemann, 2012). While a membership cutoff point suggested by existing theories or literature is ideal, such guidance for all conditions used in QCA research may be hard to come by. In such a case, QCA researchers perform cluster analysis to find thresholds for the member of each condition. For this study, the cutoff points for three explanatory conditions other than pre-existing greenspace were determined by performing the cluster analyses of raw data (see Appendix 6A.1 for raw data and Appendix 6C for the cluster analysis).

The analytical process starts with examining if any of the explanatory conditions can be a necessary condition for the outcome condition (see Appendix 6A.2). If a necessary condition is not found, we then create the truth table for the sufficiency analysis (Appendix 6A.3) and compare all possible configurations (i.e., unique sets of conditions in the table) in a stepwise way to minimize the causal paths needed to achieve the outcome. A truth table in QCA can have a maximum of 2^k unique configurations where *k* means the number of explanatory conditions for the two possible states (presence and absence) of an outcome condition. Rows in the truth table represent unique configurations and include corresponding cases that share the same configuration. Some configurations may be identified as not having empirical cases and are then referred to as "logical remainders". Logical remainders are used differently when producing three different types of solutions in QCA: *conservative* by not making assumptions about logical remainders (relies solely on empirical observations); *parsimonious*, in which any assumptions about logical remainders are allowed that lead to simpler solutions[2]; and *intermediate* by using directional expectations of some conditions based on existing theories (See Appendix 6A.4).

The model fit to assess the results produced in QCA analyses uses 'consistency' and 'coverage' scores for solution paths. The consistency score ranges from 0 to 1, which shows the degree to which the empirical data (i.e., all cases in the analysis) were consistent with the postulated relation between a solution path and the outcome condition. The more the consistency score moves from 1 to 0, the more the empirical data deviates from the sufficient relation between the solution path and the outcome condition. The coverage score represents how much of the outcome is covered by each solution path (raw coverage) or all solution paths (overall coverage) (Schneider & Wagemann, 2012). Since some cases in solution paths may overlap, unique coverage shows the difference between the raw coverage of a solution path and its intersection with other solution paths. The calculations for consistency and coverage scores are provided in the note for Table 6.3.

We analyzed the data using the QCA package in the *R* software (version 3.5.2).

[2] See Duşa (2022) for a discussion. Because assumptions used in the parsimonious case are algorithmic, not based on theory, and may use impossible counterfactuals, intermediate solutions outperform them.

A Case Study of Chicago Neighborhoods in the 2010s

Chicago was one of the seven largest cities in the US, and the seven largest cities explained almost half of the gentrification nationwide in the 2000s (Richardson et al., 2020). In addition, Chicago experienced a housing boom between 2000 and 2006. These factors make the city appropriate for the study of gentrification (Guerrieri et al., 2013). Following the 2005 "Environmental Action Agenda," the City of Chicago undertook several large-scale greening initiatives through public-private partnerships (Young, 2010). Since then, greenspace's role in predicting gentrification has increased (Stuhlmacher et al., 2022), and significant increases in housing costs have been documented around new, large park spaces (Smith et al., 2016).

Case Selection

Case selection begins with identifying Chicago community areas that are similar in geographic distance to downtown (see Fig. 6.3) and include gentrification-eligible census tracts in 2010 (GE10) and new parks inaugurated between 2008 and 2020, based on data from the Trust for Public Land. The inclusion of parks inaugurated 2 years prior to 2010 is based on the announcement effect that Immergluck and Balan (2018) reported, and this same procedure was employed by Conolly and Anguelovski (2021) in their analyses. Since park inauguration does not necessarily mean the opening of a new park, the announcement effect describes the change in investment activities that may occur after the park is announced but before it is completed and formally opened. We excluded community areas supported by the INVEST South/West project, which is a city-level neighborhood development initiative for disinvested communities on Chicago's south and west sides because we cannot discern whether

Fig. 6.3 Census tracts (CTs) selected for mv-QCA
Notes: *CA* are Community Areas; *GE10* are CTs that are Gentrification Eligible in 2010

Table 6.1 Research design and cases

	2010 GE *with* new park	2010 GE *without* new park	Total cases
Gentrified in 2020	6	9	15
Not gentrified in 2020	4	7	11
Total cases	10	16	26

neighborhood changes in these census tracts are due to the new greening activities or the large investments ($2.2. billion to date) made through INVEST South/West (City of Chicago, 2024). This led us to focus on five community areas that have 7 new parks, which include the 606 Trail in Chicago (i.e., the long horizontal park between census tracts with a green buffer at the top of Fig. 6.3), during the study period. We identified 10 gentrification-eligible census tracts in 2010 (GE10) with 7 new parks. By examining multiple cases within a single city, we hold several variables constant (such as overall changes in Chicago's economic health) to isolate the consequences of the addition of urban environmental amenities on environmental justice, as Gould and Lewis (2017) suggested.

To construct a comparison group that consists of gentrification-eligible census tracts without new parks, first, a 1-kilometer buffer was drawn from the boundary of each greenspace polygon in Fig. 6.3. The buffer represents areas that are within walking distance from a new park and may still be under the influence of a new park. Therefore, those gentrification-eligible neighborhoods in 2010 adjacent to the ones with new parks but outside the 1-km buffer were selected as a comparison group (see Fig. 6.3). Table 6.1 shows GE10 tracts with and without a new park: 10 gentrification-eligible (GE) neighborhoods in 2010 *with* a new park and 16 GE in 2010 nearby those neighborhoods *without* a new park for comparison. A total of 26 cases are included in the analysis.

Outcome Condition: Gentrification (Gent)

We identified tracts that meet the criteria for potentially experiencing gentrification in 2010 within the five community areas. Gentrification eligibility was determined by a median income lower than the average median income of the metropolitan areas (Chapple et al., 2017; Ellen & Ding, 2016; Freeman, 2009; Stuhlmacher et al., 2022).

To identify Chicago's gentrified tracts in 2020, we then examined both income and race changes between 2010 and 2020 as two separate definitions of gentrification (see Appendix 6B; Stuhlmacher et al., 2022). The class-based definition, which is measured using changes in median household income, percentage of college-educated, and housing market changes, does not capture the changes in racial composition that indicate cultural displacement (de Oliver, 2016), the loss of social capital (Imbroscio, 2008), and the loss of sense of place (Shaw & Hagemans, 2015). On the other hand, the race-based definition, which is measured using changes in non-Hispanic white or Hispanic population and housing market changes, does not capture the 'spatial invasion' of high-income gentrifiers displacing the existing working-class people (Glass, 1964).[3] In this study, the gentrification status of cases identified by the two definitions of gentrification were mostly the same (except for 2 out of 26); therefore, we used a class-based definition to define gentrified tracts in 2020. Table 6.1 shows that, of the 26 GE census tracts in 2010, 15 were gentrified and 11 were not gentrified in 2020.

Explanatory Conditions

New Park (Park)

We employ multi-value QCA (mv-QCA) to examine the three different types of urban greening on gentrification: no new park {0}, small new park {1}, and large new park {2}. The multi-value scores of set *New Park* conceptually include non-membership cases where no new parks were inaugurated. Employing the Trust for Public Land (TPL) park boundaries, we measured the size of each park in acres. Due to a lack of theoretical guidance on what defines small or large parks, we identified thresholds for exclusion, cutoff, and inclusion based on the clusters of new park sizes in the dataset. In Appendix 6C, there are several cases without a new park, two cases with less than 10 acres as a distinct cluster, and remaining cases with greater than 10 acres as another distinct cluster. Cases with values above 10.59 acres were assigned the multi-value {2} while cases with values below the size were assigned the multi-value {1}. Cases without a new park were assigned {0}. See Table 6.2 for conditions, abbreviation, measures used, and membership assignment.

Pre-existing Greenspace (Igreen)

Landsat satellite imagery was used to identify greenspaces via the Normalized Difference Vegetation Index (NDVI; Stuhlmacher et al., 2022). Identified greenspace is then superimposed with the 2010 Land Use Inventory of the Chicago Metropolitan Agency for Planning to determine existing greenspaces that are park, non-park, and vacant land (Stuhlmacher et al., 2022). Vacant lots are excluded from the calculation of total greenspace coverage because they are often considered urban disamenities because of their association with greater crime and decreased sense of safety (Deng, 2015; Foster et al., 2010; Nasar & Jones, 1997); some are

---

[3] Use of Hispanics rather than African Americans is based on the fact that the "Black population has declined in Chicago for a variety of reasons besides gentrification (Córdova et al., 2020)" (Stuhlmacher et al., 2022).

Table 6.2 Conditions, measures, and calibration for mv-QCA

Condition	Abbreviation*	Measurement	Calibration
Gentrification	Gent	2010 GE gentrified in 2020	GE not gentrified in 2020 = 0, GE gentrified in 2020 = 1
New park	Park	Presence and acreage of a new park in the 2010s	No Park = 0, Small Park = 1 (< 10.59 acres), Large Park = 2 (10.59 acres or larger)
Pre-existing greenspace	Igreen	Greenspace coverage per person in 2010	Insufficient = 0 (< 9 m^2/person) Sufficient = 1 (≥9 m^2/person)
Government-assisted housing units	Ghous	Percentage of affordable rental housing units in the community area	Relatively low = 0 (< 1.04%), Relatively high = 1 (1.04% or higher)
Economic spillover	Gneigh	The total size of neighboring census tracts gentrified between 2000 and 2010	Absence = 0 (< 1140 km^2), Presence = 1 (1140 km^2 or larger)

Notes. See Appendix 6A.1 for raw data; When we refer to the name of conditions in this chapter, we capitalized the first letter of a condition and lowered subsequent letters (e.g., Park for the new park condition). When all letters of a condition are written as upper-case letters, it represents the presence of the condition (e.g., PARK). Lower-case letters are used for the absence of the condition (e.g., park)

brownfields. The area coverage of park and non-park greenspaces are summed within each census tract and divided by each census tract population in 2010 to determine whether the allocation of existing greenspace per person in the census tract meets the minimum recommended level. The recommended coverage of minimum greenspace per person is 9 m^2 (WHO, 2012); therefore, a cutoff value for the membership score of adequate pre-existing greenspace per person (IGREEN) is set at 9 m^2.

Government-Assisted Housing Units (Ghous)

Affordable housing initiatives supported by the Chicago Department of Housing provide affordable rental housing opportunities to low-income individuals and families, including seniors, veterans, and the homeless. These housing units are built, renovated, and maintained with the help of the city of Chicago. The local government's community development efforts are based on the community boundaries (as in projects like INVEST South/West); therefore, the number of affordable rental units per community area is divided by the total housing units in 2010 within the community area. The Near West Side community area has the highest percentage of affordable rental housing units, about 6% of the total housing units. To assign cases for the membership of relatively higher government-assisted housing (GHOUS), we performed cluster analysis of raw data values as we have done for new park condition. The cutoff to be a member of this condition is 1.04% of all housing in the community area being affordable housing.

Economic Spillover (Gneigh)

To operationalize whether a neighborhood is under a considerable economic spillover effect, we first identified if adjacent census tracts that share edges and corners of each census tract were gentrified in the previous decade (i.e., between 2000 and 2010). We then use the total square meters of those gentrifying neighbors to determine whether the census tract is with or without considerable gentrifying neighbors. To account for the irregular shape of the census boundary, we use the total size of gentrifying neighbors instead of the number of gentrifying neighbors. The total size indicates whether a census tract is under a considerable economic spillover effect of its gentrifying neighbors. The inclusion of membership in the GNEIGH condition implies that a census tract is near gentrifying neighbors and is under the considerable influence of the gentrification spillover effect. The cutoff to be a member of the condition is at 1140 km^2, which means that if the size of gentrifying neighbors is greater than the cutoff (1140 km^2), a census tract is considered under economic spillover (GNEIGH).

Findings

None of the park types were a necessary condition for gentrification (see Appendix 6A.2): there are census tracts where gentrification occurred even if new greening was not introduced to a neighborhood. Given the lack of findings from the necessity analysis, here we focus on explaining the combination of causal conditions that are *sufficient* to explain the outcome conditions. We report and interpret the intermediate solution below because there were no great differences between conservative and intermediate solutions (see the conservative and parsimonious solutions in Appendix 6A.4).

Table 6.3 shows three sufficient causal paths to green gentrification. They draw a holistic picture of gentrification scenarios within different conditions of new greening; there are multiple pathways. All three different types of greening conditions – large new park, small new park, and no new park– can lead to gentrification in conjunction with specific neighborhood conditions. The findings of the three pathways can then be thought of as complementing information for building a deeper understanding of each condition in relation to gentrification. Equation 6.1 provides the QCA solution expression:

Table 6.3 Paths to gentrification

Intermediate solution	Consistency	Raw coverage	Unique coverage	Neighborhood ID[a]	Mechanism
Path1: LPARK*igreen *GHOUS	0.833	0.333	0.333	1,4,5,6,7; 3	"Green Growth Machine"
Path 2: SPARK*igreen*ghous	1.000	0.067	0.067	9	"Small-Park-involved Gentrification"
Path 3: NOPARK*igreen *GHOUS*GNEIGH	0.800	0.267	0.267	11,12,13,14, 18	"Spillover Gentrification"
Overall fit	0.833	0.667			

Note. LPARK = Large Park {2}; SPARK = Small Park {1}; NOPARK = No Park {0}; A semicolon in the Neighborhood ID column differentiates cases in different configurations (see Appendix 6A.3)
The parameters of fit are calculated using the following equations

$$Consistency_{sufficient\,conditions(x_i \le y_i)} = \frac{\sum_{I}^{i=1} \min(X_i, Y_i)}{\sum_{I}^{i=1} X_i}$$

$$Coverage_{sufficient\,conditions(x_i \le y_i)} = \frac{\sum_{I}^{i=1} \min(X_i, Y_i)}{\sum_{I}^{i=1} Y_i}$$

where *Y* is the outcome condition, *X* can be a condition, a solution path, or solution paths, and *i* indicates a case. The *min* is the lowest value of *X* and *Y*. The unique coverage score is calculated by subtracting the coverage of the intersection between path 1, path 2, and path 3 from the raw coverage of path 1 (and so forth, depending on the path under consideration)
[a]See Appendix 6A for each neighborhood ID

$$\left(\text{LPARK}^{*}\text{igreen}^{*}\text{GHOUS}\right) + \left(\text{SPARK}^{*}\text{igreen}^{*}\text{ghous}\right) + \left(\text{NOPARK}^{*}\text{igreen}^{*}\text{GHOUS}^{*}\text{GNEIGH}\right) \rightarrow \text{GENT} \tag{6.1}$$

Note that upper-case letters represent the presence of a condition and lower-case letters are used for the absence of a condition. Conditions combined with logical AND (*) show conjunctional causes that are sufficient to produce gentrification. These multiple paths are sufficient conditions that can be separately expressed using logical OR (+) to explain the outcome. The arrow to the right (→) implies a sufficiency relationship between the configurations and the outcome condition in QCA. The expression can also be presented as in Table 6.3.

*Path 1 "Green Growth Machine": LPARK*igreen*GHOUS → GENT*

The first path illustrates a configuration where a large new park contributed to gentrification. As opposed to the proposal of making housing affordable as an anti-gentrification measure, a relatively higher level of government-assisted housing units in a community area was not enough to deter the gentrification effect of a large new park when there is inadequate existing greenspace per person. The mechanism manifested by the first path confirms what scholars call the "Green Growth Machine," which describes the urban redevelopment driven both by public and private interests around ecological restoration (Rigolon & Nemeth, 2018; Young, 2010). The cases covered in this path are neighborhoods in Logan Square that saw the inauguration of the 606 Trail in 2015. The 606 or Bloomingdale Trail was a major railway restoration project that was advocated by community members for several years but taken up and pushed through by Mayor Rahm Emanual's administration with considerable private fundraising conducted by the Trust for Public Lands (Rigolon & Németh, 2018).

*Path 2 "Small-Park-Involved Gentrification": SPARK*igreen*ghous → GENT*

The second path depicts an alternative in which a small new park still contributes to gentrification. This combination suggests a relatively lower level of government-assisted housing units in a community area along with insufficient existing greenspace per person is enough to encourage gentrification even when a new park is small. The path includes a new greening that is not large enough to bring about the "Green Growth Machine" mechanism induced by large-scale urban redevelopment efforts. When compared to the first path's combination of a large new park and a relatively higher level of government-assisted housing units, the second path implies the importance of the level of government-assisted housing units within the context of park size for the different gentrification outcomes. In other words, both the insufficiency of existing greenspace and government's relative attention to affordable housing in a community area are important contextual factors when considering the gentrification outcome of a small new park.

*Path 3 "Spillover Gentrification": NOPARK*igreen*GHOUS *GNEIGH → GENT*

The third path includes gentrification with no new parks contributing. When compared with the first path, the third path differs in park type and the presence of economic spillover. Even without a new park, neighborhoods with insufficient greenspace but in the vicinity of neighborhoods gentrifying in the previous decades were gentrified. Similar to the first path, a relatively higher level of government-assisted housing units was not enough to deter gentrification, which signifies the strength of the force of economic development in nearby areas. Four out of five neighborhoods covered in the third path (ID 11, 12, 13, 14; i.e., cases without a new park) are all in the same community area (Logan Square) as are four out of six neighborhoods covered in the first path (ID 1, 3, 4, 5; i.e., cases with a new park). This community area experienced intense neighborhood improvement (c.f., Harris et al., 2021), illustrating that the forces of economic and physical development might be the underlying condition for gentrification. Thus, this path is called "Spillover Gentrification."

Discussion

This study examines multiple neighborhoods in Chicago to find combinations of causal conditions that are sufficient to explain gentrification with and without new urban greening. Three sufficient causal paths were found, including a path without a new park. These findings indicate that a new greening project may not be the decisive factor for some neighborhoods to experience gentrification. The implication of these findings is critical for two reasons. First, the past discourse on green gentrification literature might have been overemphasizing the gentrification effect of new greening, especially due to the concern placed on displacement by the environmental justice point of view. Second, green gentrification literature should address the elephant in the room: sorting out the temporal and dependent relationship between new greening and large-scale urban redevelopment efforts in explaining gentrification.

The focus of green gentrification literature is moving toward finding more empirical evidence by including more cities in the analysis and finding the greater influence of new greening on gentrification (e.g., Anguelovski et al., 2022). While the green gentrification literature raises invaluable awareness toward environmental justice issues around new greening initiatives among city officials and practitioners, there is a lack of attention to alternative ways to manage green gentrification. For example, the potential unintended consequences of urban greening, despite its diverse environmental and health benefits, create fear of displacement among low-income residents when the local government expresses its intention to invest in urban greenspace in their neighborhoods. Yet, ironically, residents who are concerned with the gentrification effect of a new greening initiative, and their dissenting voices to park projects, are often disregarded in the name of urban greening as a universal good (Fernandez et al., 2021). The overly emphasized role of greening in the discussion of gentrification, without noting that avoiding greening may not prevent gentrification, and without providing any alternative ways to achieve greenspace justice, will only intensify the tension between greening advocacy and social justice.

Recently, scholars have begun to recognize the need for strategies and tools to help communities achieve just outcomes from urban greening (Rigolon et al., 2020); yet, these efforts are operating under the assumption that urban greening is a decisive factor that drives gentrification. Our findings show otherwise: that neighborhoods without a new park similar to the ones with a new park could still be gentrified, especially under the influence of economic development in the nearby areas. This finding not only indicates that new greening initiatives may not be the decisive factor driving gentrification, but also guides our understanding of gentrification in conjunctional ways to approach urban greening initiatives based on different local contexts (e.g., levels of pre-existing greenspace). Contextualized and localized understanding of gentrification has the potential to help devise ways to alleviate the gentrification effect of new greening and ease the tension between diverse stakeholders in the community to better communicate about a new greening project.

When compared to the "Green Growth Machine" path, which describes gentrification encouraged by a large new park, the "Spillover Gentrification" path, which describes the economic spillover being a differentiating factor causing neighborhoods without a new park to experience gentrification, highlights a potential issue of cyclical causal direction between new greening and gentrification (Rigolon & Collins, 2022). These two paths with and without a new park allow us to posit that a large new park may represent the intention of community-wide economic development that includes urban greening as one component. The causal direction, in this case, is the opposite of what green gentrification literature has been arguing. In other words, in some cases, neighborhood development may precede urban greening, not the greening preceding and encouraging gentrification.

Sorting out the complex relationship between urban greening and gentrification is analytically challenging. Researchers sought to resolve the issue through the a priori designation of gentrifiable tracts that could signal underdevelopment (Quinton et al., 2022). However, as discussed in Chap. 4, this practice fails to account for the notion of gentrification being an ongoing process rather than a definitive phenomenon with a specific start and endpoint. Nonetheless, challenges remain to capture the not-yet materialized but intended investment activities and large-scale urban redevelopment efforts. If greening and development efforts are linked and integrated as Anguelovski et al. (2022) have found in some cities, we need a more refined analytical approach to separate them out to identify when or whether urban greening is a decisive gentrifying factor that creates environmental injustice.

Indeed, these nuanced, refined understandings of green gentrification are more possible when a different methodological approach is combined with a new research design. Combining the set-theoretic method with a comparison-group design offers an alternative way to examine whether new urban greening is the difference-maker for neighborhoods experiencing gentrification. The set-theoretic method allows us to take a stance grounded in the perspective that the meaning of greenup categories, such as a new park, is constructed based on the socially agreed languages and thoughts specific to time and space (Mahoney, 2021). In this study, we take relevant gentrification factors, including the addition of a new park, as concepts that are constructed and measured based on the socially agreed inclusion and exclusion of a set (i.e., membership status in a concept) and explore causal pathways that emerged from their set relationships in a given temporal sequence (Mahoney, 2021:46). By adding a comparison group in the set-theoretic method, we are able to compare emerged causal pathways to infer under what conditions a specific concept (e.g., the addition of a new park) is an inevitable component for the outcome set to occur. Future studies of green gentrification may benefit from employing the set-theoretic method with a comparison-group design to further narrow the methodological gap in the literature and deepen our understanding of the phenomenon to suggest equitable ways to provide urban greening in a given context.

Conclusion

We employed a QCA approach to refine our understanding of the gentrification effect of new greening initiatives in Chicago in the 2010s. Our study suggests that new green investment may not be a decisive factor for some neighborhoods to experience gentrification; in some cases gentrification occurs without a new green amenity, and gentrification does not always occur with one. When trying to reduce gentrification, we recommend giving careful attention to the initiation of a small new park in a community area where government-assisted affordable housing units are relatively low because a small new park may encourage gentrification in such a neighborhood. And, even without a new park, neighborhoods can experience gentrification when nearby areas have begun gentrifying in the previous decades.

Rather than overemphasizing the potential gentrification effect of new green investment and intensifying the tension between environmental advocacy and social justice, scholars of green gentrification should pay attention to the analytical challenge of sorting out the temporal and dependent relationship between new green investment and large-scale urban redevelopment efforts. If such efforts are materialized, we may have a better understanding of the different justice outcomes of different types of new green investments.

Appendixes

Appendix 6A.1 Raw data

Neighborhood ID	Community Area	Gent	Park		Igreen		Ghous		Gneigh	
			acre	multi-value (mv) score	m^2	Membership	%	Membership	km^2	Membership
1	Logan Square	1	16.76	(2)	1.59	(0)	1.67	(1)	496.69	(0)
2	Logan Square	0	17.27	(2)	9.1	(1)	1.67	(1)	711.47	(1)
3	Logan Square	1	16.76	(2)	1.19	(0)	1.67	(1)	711.47	(1)
4	Logan Square	1	16.76	(2)	1.21	(0)	1.67	(1)	214.78	(0)
5	Logan Square	0	16.76	(2)	4.86	(0)	1.67	(1)	384.22	(0)
6	West Town	1	16.76	(2)	0	(0)	1.48	(1)	496.2	(0)
7	West Town	1	16.76	(2)	1.38	(0)	1.48	(1)	166.27	(0)
8	Near West Side	0	2.21	(1)	0	(0)	6.02	(1)	439.3	(0)
9	Bridgeport	1	4.41	(1)	0	(0)	0.6	(0)	0	(0)
10	South Lawndale	0	22.77	(2)	12.89	(1)	0.19	(0)	0	(0)
11	Logan Square	0	0	(0)	4.45	(0)	1.67	(1)	714.53	(1)
12	Logan Square	1	0	(0)	0	(0)	1.67	(1)	785.81	(1)
13	Logan Square	1	0	(0)	3.38	(0)	1.67	(1)	1319.27	(1)
14	Logan Square	1	0	(0)	4.57	(0)	1.67	(1)	665.35	(1)
15	Logan Square	1	0	(0)	0	(0)	1.67	(1)	0	(0)
16	Logan Square	1	0	(0)	1.19	(0)	1.67	(1)	0	(0)
17	West Town	1	0	(0)	521.8	(1)	1.48	(1)	1308.07	(1)

(continued)

Appendix 6A.1 (continued)

Neighborhood ID	Community Area	Gent	Park		Igreen		Ghous		Gneigh	
			acre	multi-value (mv) score	m^2	Membership	%	Membership	km^2	Membership
18	West Town	1	0	(0)	0.09	(0)	1.48	(1)	971.94	(1)
19	South Lawndale	0	0	(0)	0.43	(0)	0.19	(0)	0	(0)
20	South Lawndale	0	0	(0)	1.91	(0)	0.19	(0)	0	(0)
21	South Lawndale	0	0	(0)	0	(0)	0.19	(0)	0	(0)
22	Bridgeport	1	0	(0)	2.06	(0)	0.6	(0)	0	(0)
23	Near West Side	0	0	(0)	5.75	(0)	6.02	(1)	439.3	(0)
24	South Lawndale	1	0	(0)	7.82	(0)	0.19	(0)	0	(0)
25	South Lawndale	0	0	(0)	10.87	(1)	0.19	(0)	0	(0)
26	Near West Side	0	0	(0)	18.18	(1)	6.02	(1)	1926.29	(1)

Appendix 6A.2 Necessity analysis of park types

		Consistency	Relevance
1	LPARK	0.333	0.857
2	SPARK	0.067	0.960
3	NOPARK	0.600	0.588

Notes: A condition deemed as necessary meets two criteria in the parameters of fit: the consistency threshold of 0.9 (as advised by Ragin, 2006) and the relevance threshold of 0.6 (Duşa, 2018, p. 123). Relevance score is the coverage of a necessity condition which shows how closely overlapped set X and set Y values are. It is calculated using the following formula:

$$Relevance_{necessary conditions(x_i \geq y_i)} = \frac{\sum_{I}^{i=1}(1 - X_i)}{\sum_{I}^{i=1}(1 - \min(X_i, Y_i))}$$

where Y is the outcome condition, X can be a condition, a solution path, or solution paths, and i indicates a case. The *min* is the lowest value of X and Y

Appendix 6A.3 Truth table for green gentrification

Row	Park	Igreen	Ghous	Gneigh	OUT	n	Incl.	Neighborhood ID
1	1	0	0	0	1	1	1.000	9
2	2	0	1	1	1	1	1.000	3
3	0	0	1	1	1	5	0.800	11, 12, 13, 14, 18
4	2	0	1	0	1	5	0.800	1, 4, 5, 6, 7
5	0	0	1	0	0	3	0.667	15, 16, 23
6	0	1	1	1	0	2	0.500	17, 26
7	0	0	0	0	0	5	0.400	19, 20, 21, 22, 24
8	0	1	0	0	0	1	0.000	25
9	1	0	1	0	0	1	0.000	8
10	2	1	0	0	0	1	0.000	10
11	2	1	1	1	0	1	0.000	2

See Table 6.2 for the definition of conditions: Park, Ghous, Igreen, and Gneigh
OUT: outcome value to be included (1) or not (0) for logical minimization based on the inclusion score of greater than 0.75
n number of cases (neighborhoods) in configuration; *Incl.* indicates sufficiency inclusion score (interchangeably used as consistency score)

Appendix 6A.4 Other solutions

Conservative solution	Consistency	Raw coverage	Unique coverage	Neighborhood ID
LPARK*igreen*GHOUS	0.833	0.333	0.333	1,4,5,6,7; 3
SPARK*igreen*ghous*gneigh	1.000	0.067	0.067	9
NOPARK*igreen*GHOUS	0.750	0.400	0.400	11,12,13,14, 18; 15,16,23
Overall fit	0.800	0.800		

Parsimonious solution		Consistency	Raw Coverage	Unique Coverage	Neighborhood ID
M1	SPARK*ghous	1.000	0.067	0.067	9
	igreen*GNEIGH	0.833	0.333	0.267	11,12,13,14,18; 3
Overall fit		0.833	0.667		
M2	LPARK*igreen	0.833	0.333	0.000	1,4,5,6,7; 3
	LPARK*GHOUS *gneigh	0.800	0.267	0.000	1,4,5,6,7
Overall fit		0.833	0.667		

Appendix 6B Defining gentrification

Class-based gentrification	Race-based gentrification
1. Gentrification eligible tracts (GE) in 2010 a. Median income of a tract lower than the citywide average median income	1. Gentrification eligible tracts (GE) in 2010 a. Median income of a tract lower than the citywide average median income
2. Demographic changes between 2010 and 2020: a. Positive change in median household income > average city change, and b. Positive change in % college educated > average city change	2. Demographic changes between 2010 and 2020: a. Increase of % non-Hispanic white population > 1 standard deviation above the mean, or b. Decrease of % Hispanic population > 1 standard deviation above the mean
3. Housing market changes between 2010 and 2020: a. Increase in median gross rent greater than average city change, or b. Increase in median home value greater than average city change	3. Housing market changes between 2010 and 2020: a. Increase in median gross rent, greater than average city change, or b. Increase in median home value greater than average city change

Note: To be identified as a gentrified neighborhood in 2020, census tracts must satisfy the three conditions listed in the column that defines class-based or race-based gentrification (Stuhlmacher et al., 2022)

Appendix 6C Membership assignment based on the cluster analysis of *New Park*

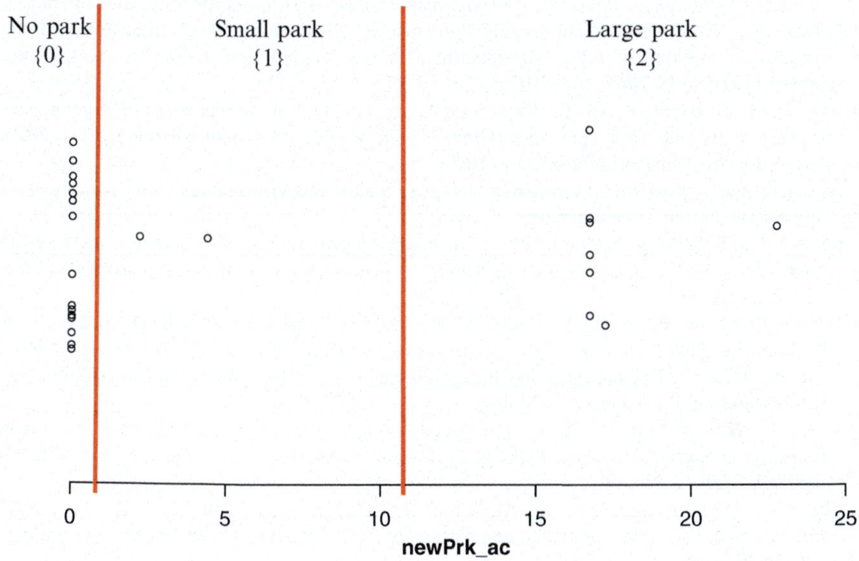

References

Anguelovski, I., Connolly, J. J. T., Masip, L., & Pearsall, H. (2018). Assessing green gentrification in historically disenfranchised neighborhoods: A longitudinal and spatial analysis of Barcelona. *Urban Geography, 39*(3), 458–491.

Anguelovski, I., Connolly, J. J., Garcia-Lamarca, M., Cole, H., & Pearsall, H. (2019). New scholarly pathways on green gentrification: What does the urban 'green turn' mean and where is it going? *Progress in Human Geography, 43*(6), 1064–1086. https://doi.org/10.1177/0309132518803799

Anguelovski, I., Connolly, J. J. T., Cole, H., Garcia-Lamarca, M., Triguero-Mas, M., Baró, F., Martin, N., Conesa, D., Shokry, G., del Pulgar, C. P., Ramos, L. A., Matheney, A., Gallez, E., Oscilowicz, E., Máñez, J. L., Sarzo, B., Beltrán, M. A., & Minaya, J. M. (2022). Green gentrification in European and North American cities. *Nature Communications, 13*(1), 3816. https://doi.org/10.1038/s41467-022-31572-1

Bai, H., Stanis, S. A. W., Kaczynski, A. T., & Besenyi, G. M. (2013). Perceptions of neighborhood park quality: Associations with physical activity and body mass index. *Annals of Behavioral Medicine, 45*(suppl 1), S39–S48. https://doi.org/10.1007/S12160-012-9448-4

Bona, S., Silva-Afonso, A., Gomes, R., Matos, R., & Rodrigues, F. (2023). Nature-based solutions in urban areas: A European analysis. *Applied Sciences, 13*(1), 1. https://doi.org/10.3390/app13010168

Brazil, N., & Portier, A. (2022). Investing in gentrification: The eligibility of gentrifying neighborhoods for federal place-based economic investment in US cities. *Urban Affairs Review, 58*(5), 1234–1276.

Bryson, J. (2012). Brownfields gentrification: Redevelopment planning and environmental justice in Spokane, Washington. *Environmental Justice, 5*(1), 26–31. https://doi.org/10.1089/env.2010.0045

Chapple, K., Loukaitou-Sideris, A., Waddell, P., Chatman, D., Ong, P., & Zuk, M. (2017). *Developing a new methodology for analyzing potential displacement*. California Air Resources Board.
Checker, M. (2011). Wiped out by the "greenwave": Environmental gentrification and the paradoxical politics of urban sustainability. *City & Society, 23*(2), 210–229. https://doi.org/10.1111/j.1548-744X.2011.01063.x
Chen, Y., Xu, Z., Byrne, J., Xu, T., Wang, S., & Wu, J. (2021). Can smaller parks limit green gentrification? Insights from Hangzhou, China. *Urban Forestry & Urban Greening, 59*, 127009. https://doi.org/10.1016/j.ufug.2021.127009
City of Chicago. (2024). *Invest South/West*. INVEST South/West Home. https://www.chicago.gov/city/en/sites/invest_sw/home.html
Connolly, J. J. T., & Anguelovski, I. (2021). Three histories of greening and whiteness in American cities. *Frontiers in Ecology and Evolution, 9*. https://www.frontiersin.org/article/10.3389/fevo.2021.621783
Córdova, T., Ewing, E., Lee, L.Y., Loury, A., Pattillo, M., Ransby, B., Stovall, D., Sutton, S., & William, S. (2020). *Between the great migration and growing exodus: The future of black Chicago?* State of Racial Justice in Chicago Project. University of Illinois at Chicago, Institute for Research on Race and Public Policy.
Curran, W., & Hamilton, T. (2012). Just green enough: Contesting environmental gentrification in Greenpoint, Brooklyn. *Local Environment, 17*(9), 1027–1042. https://doi.org/10.1080/13549839.2012.729569
Curran, W., Stuhlmacher, M., & Anderson, E. (2023). Getting to just green enough. *Interdisciplinary Journal of Landscape Architecture (LA+), BEAUTY, 18*. https://laplusjournal.com/18-BEAUTY-1
de Oliver, M. (2016). Gentrification as the appropriation of therapeutic 'diversity': A model and case study of the multicultural amenity of contemporary urban renewal. *Urban Studies, 53*(6), 1299–1316. https://doi.org/10.1177/0042098015576314
Deng, C. (2015). Integrating multi-source remotely sensed datasets to examine the impact of tree height and pattern information on crimes in Milwaukee, Wisconsin. *Applied Geography, 65*, 38–48.
Derickson, K., Klein, M., & Keeler, B. L. (2021). Reflections on crafting a policy toolkit for equitable green infrastructure. *Npj Urban Sustainability, 1*(1), 21. https://doi.org/10.1038/s42949-021-00014-0
Duşa, A. (2018). *QCA with R: A comprehensive resource*. Springer.
Duşa, A. (2022). Critical tension: Sufficiency and parsimony in QCA. *Sociological Methods & Research, 51*(2), 541–565.
Easton, S., Lees, L., Hubbard, P., & Tate, N. (2020). Measuring and mapping displacement: The problem of quantification in the battle against gentrification. *Urban Studies, 57*(2), 286–306. https://doi.org/10.1177/0042098019851953
Ellen, I. G., & Ding, L. (2016). Advancing our understanding of gentrification. *Cityscape: A Journal of Policy Development and Research, 18*(3), 3–7.
Fernandez, M., Harris, B., & Rose, J. (2021). Greensplaining environmental justice: A narrative of race, ethnicity, and justice in urban greenspace development. *Journal of Race, Ethnicity and the City, 2*(2), 210–231. https://doi.org/10.1080/26884674.2021.1921634
Foster, S., Giles-Corti, B., & Knuiman, M. (2010). Neighbourhood design and fear of crime: A social-ecological examination of the correlates of residents' fear in new suburban housing developments. *Health & Place, 16*, 1156–1165.
Freeman, L. (2009). Neighbourhood diversity, metropolitan segregation and gentrification: What are the links in the US? *Urban Studies, 46*(10), 2079–2101.
Glass, R. (1964). *London: Aspects of change*. MacGibbon & Kee.
Goertz, G. (2003). Assessing the importance of necessary or sufficient conditions in fuzzy-set social science. *COMPASSS Working Papers, 2003*(7). http://www.compasss.org/wpseries/Goertz2003.pdf

Gould, K., & Lewis, T. (2017). *Green gentrification: Urban sustainability and the struggle for environmental justice*. Routledge. https://doi.org/10.4324/9781315687322

Guerrieri, V., Hartley, D., & Hurst, E. (2013). Endogenous gentrification and housing price dynamics. *Journal of Public Economics, 100*, 45–60. https://doi.org/10.1016/j.jpubeco.2013.02.001

Harris, B., Schmalz, D., Larson, L., & Fernandez, M. (2021). Fear of the unknown: Examining neighborhood stigma's effect on urban greenway use and surrounding communities. *Urban Affairs Review, 57*(4), 1015–1048. https://doi.org/10.1177/1078087420909529

Imbroscio, D. (2008). "[U]nited and actuated by some common impulse of passion": Challenging the dispersal consensus in American housing policy research. *Journal of Urban Affairs, 30*(2), 111–130.

Immergluck, D., & Balan, T. (2018). Sustainable for whom? Green urban development, environmental gentrification, and the Atlanta Beltline. *Urban Geography, 39*(4), 546–562. https://doi.org/10.1080/02723638.2017.1360041

Kabisch, N., Frantzeskaki, N., Pauleit, S., Naumann, S., Davis, M., Artmann, M., Haase, D., Knapp, S., Korn, H., Stadler, J., Zaunberger, K., & Bonn, A. (2016). Nature-based solutions to climate change mitigation and adaptation in urban areas: Perspectives on indicators, knowledge gaps, barriers, and opportunities for action. *Ecology and Society, 21*(2) https://www.jstor.org/stable/26270403

Kim, J., Kim, Y., & Stuhlmacher, M. (2024). The green space dilemma: Pathways to greening with and without gentrification. *Journal of Urban Affairs*.

Kwon, Y., Joo, S., Han, S., & Park, C. (2017). Mapping the distribution pattern of gentrification near urban parks in the case of Gyeongui Line Forest Park, Seoul, Korea. *Sustainability, 9*, 231.

Loughran, K. (2014). Parks for profit: The high line, growth machines, and the uneven development of urban public spaces. *City & Community, 13*(1), 49–68.

Mabon, L., & Shih, W.-Y. (2021). Urban greenspace as a climate change adaptation strategy for subtropical Asian cities: A comparative study across cities in three countries. *Global Environmental Change, 68*, 102248. https://doi.org/10.1016/j.gloenvcha.2021.102248

Mahoney, J. (2021). *The logic of social science*. Princeton University Press.

McCormack, G. R., Rock, M., Toohey, A. M., & Hignell, D. (2010). Characteristics of urban parks associated with park use and physical activity: A review of qualitative research. *Health and Place, 16*(4), 712–726. https://doi.org/10.1016/j.healthplace.2010.03.003

Mell, I. C., Henneberry, J., Hehl-Lange, S., & Keskin, B. (2013). Promoting urban greening: Valuing the development of green infrastructure investments in the urban core of Manchester, UK. *Urban Forestry & Urban Greening, 12*(3), 296–306.

Nasar, J. L., & Jones, K. M. (1997). Landscapes of fear and stress. *Environment and Behavior, 29*, 291–323.

Oakley, D., & Burchfield, K. (2009). Out of the projects, still in the hood: The spatial constraints on public-housing residents' relocation in Chicago. *Journal of Urban Affairs, 31*(5), 589–614. https://doi.org/10.1111/j.1467-9906.2009.00454.x

Pearsall, H. (2010). From brown to green? Assessing social vulnerability to environmental gentrification in New York City. *Environment and Planning C: Government and Policy, 28*(5), 872–886. https://doi.org/10.1068/c08126

Quinton, J., Nesbitt, L., & Sax, D. (2022). How well do we know green gentrification? A systematic review of the methods. *Progress in Human Geography, 030913252211044*. https://doi.org/10.1177/03091325221104478

Ragin, C. C. (2006). Set relations in social research: Evaluating their consistency and coverage. *Political Analysis, 14*(3), 291–310.

Richardson, J., Mitchell, B., & Edlebi, J. (2020). *Gentrification and Disinvestment 2020: Do Opportunity Zones benefit or gentrify low-income neighborhoods?* National Community Reinvestment Coalition. https://ncrc.org/wp-content/uploads/2020/06/Gentrification-and-Opportunity-Zones-2020-v7.pdf

Rigolon, A., & Collins, T. (2022). The green gentrification cycle. *Urban Studies*, 00420980221114952. https://doi.org/10.1177/00420980221114952.

Rigolon, A., & Németh, J. (2018). "We're not in the business of housing:" environmental gentrification and the nonprofitization of green infrastructure projects. *Cities, 81*, 71–80. https://doi.org/10.1016/j.cities.2018.03.016

Rigolon, A., & Németh, J. (2020). Green gentrification or 'just green enough': Do park location, size and function affect whether a place gentrifies or not? *Urban Studies, 57*(2), 402–420. https://doi.org/10.1177/0042098019849380

Rigolon, A., Keith, S. J., Harris, B., Mullenbach, L. E., Larson, L. R., & Rushing, J. (2020). More than "just green enough": Helping park professionals achieve equitable greening and limit environmental gentrification. *Journal of Park & Recreation Administration, 38*(3), 29–54.

Schneider, C. Q., & Wagemann, C. (2012). *Set-theoretic methods for the social sciences: A guide to qualitative comparative analysis*. Cambridge University Press.

Sharifi, F., Nygaard, A., Stone, W. M., & Levin, I. (2021). Green gentrification or gentrified greening: Metropolitan Melbourne. *Land Use Policy, 108*, 105577. https://doi.org/10.1016/j.landusepol.2021.105577

Shaw, K. S., & Hagemans, I. W. (2015). 'Gentrification without displacement' and the consequent loss of place: The effects of class transition on low-income residents of secure housing in gentrifying areas. *International Journal of Urban and Regional Research, 39*(2), 323–341. https://doi.org/10.1111/1468-2427.12164

Smith, G., Duda, S., Lee, J. M., & Thompson, M. (2016). *Measuring the impact of the 606: Understanding how a large public investment impacted the surrounding housing market*. Institute for Housing Studies, DePaul University.

Stuhlmacher, M., Kim, Y., & Kim, J. E. (2022). The role of green space in Chicago's gentrification. *Urban Forestry & Urban Greening, 71*, 127569. https://doi.org/10.1016/j.ufug.2022.127569

Sugiyama, T., Giles-Corti, B., Summers, J., du Toit, L., Leslie, E., & Owen, N. (2013). Initiating and maintaining recreational walking: A longitudinal study on the influence of neighborhood green space. *Preventive Medicine, 57*(3), 178–182.

Triguero-Mas, M., Anguelovski, I., Connolly, J. J. T., Martin, N., Matheney, A., Cole, H. V., Pérez-del-Pulgar, C., Garcia-Lamarca, M., Shokry, G., Argüelles, L., Conesa, D., Gallez, E., Sarzo, B., Beltran, M. A., López-Mañez, J., Martínez-Minaya, J., Oscilowicz, E., Arcaya, M. C., & Baró, F. (2022). Exploring green gentrification in 28 global north cities: The role of urban parks and other types of greenspace. *Environmental Research Letters*. https://doi.org/10.1088/1748-9326/ac9325

Verweij, S. (2015). Once the shovel hits the ground: Evaluating the management of complex implementation processes of public-private partnership infrastructure projects with qualitative comparative analysis. *SSRN Electronic Journal*. https://doi.org/10.2139/ssrn.2646785

Wolch, J. R., Byrne, J., & Newell, J. P. (2014). Urban green space, public health, and environmental justice: The challenge of making cities 'just green enough'. *Landscape and Urban Planning, 125*, 234–244. https://doi.org/10.1016/j.landurbplan.2014.01.017

World Health Organization (WHO). (2012). *Health indicators of sustainable cities in the context of the Rio +20 UN Conference on Sustainable Development*. WHO.

Young, R. F. (2010). The greening of Chicago: Environmental leaders and organisational learning in the transition toward a sustainable metropolitan region. *Journal of Environmental Planning and Management, 53*(8), 1051–1068. https://doi.org/10.1080/09640568.2010.508948

Zuk, M., Bierbaum, A. H., Chapple, K., Gorska, K., & Loukaitou-Sideris, A. (2018). Gentrification, displacement, and the role of public investment. *Journal of Planning Literature, 33*(1), 31–44. https://doi.org/10.1177/0885412217716439

Chapter 7
Simulating the Consequences of Policies That Improve Environmental Conditions

Abstract This chapter explores the implications of policies that improve environmental conditions in communities. Using an agent-based model, we implement a series of different policy scenarios that site parks and recreational facilities of different sorts in different types of community settings. We find that each unique recipe of strategies results in different outcomes, but crucially find that some scenarios, particularly those most associated with the just green enough (JGE) approach, seem to slow down the gentrification and displacement process considerably.

Introduction

As detailed in earlier chapters, environmental gentrification, or green gentrification, occurs when environmental improvement in a neighborhood alters the demand for housing in a neighborhood. With improved environmental conditions, people with more wealth than the current residents of a neighborhood may see homes in the neighborhood as more appealing than before, either as potential new residences or as investment opportunities. Either of these outcomes raises prices and may drive out those living there at the time of improvement. Because of this potential, current residents in a neighborhood may be wary of environmental improvement, and city policymakers may be skeptical about making environmental improvements in neighborhoods that need such improvement if they believe that the benefits of the improvement may be enjoyed by incoming residents rather than current residents.

However, the relationship between environmental improvement and gentrification is likely much more nuanced than this, and cities may be able to deliver environmental improvements to neighborhoods if they better connect the context of the neighborhood with the improvement to be implemented. Much of the environmental gentrification literature investigates environmental improvement as the remediation of an environmental "bad," but recent literature has also explored how placing environmental goods, such as parks and greenspace, may foster gentrification trends. Thus, the concept includes both "cleaning and greening." In this chapter, complementary to Chap. 6, which used QCA to analyze past outcomes in Chicago,

H. E. Campbell et al., *Green Gentrification and Environmental Injustice*,
https://doi.org/10.1007/978-3-031-65100-7_7

we use Agent-Based Modeling (ABM) to forecast conditions under which placing new public goods, or amenities, are more and less likely to lead to green gentrification and to displacement. We also examine duration questions: If it is inevitable that some people will be displaced, are there circumstances in which this displacement plays out over a longer period of time, enabling those at risk to be displaced to remain in their community longer and therefore to benefit from the environmental improvement?

Green Gentrification

The process of green gentrification is straightforward and a direct offshoot of gentrification more generally. Green gentrification occurs when the environmental conditions in a neighborhood improve and the neighborhood then becomes more desirable for higher-wealth residents to either move in themselves or purchase properties for investment opportunities (Gould & Lewis, 2017). As these newcomers and investors purchase properties, property values increase, which increases property taxes for owners and (at least eventually) rental costs for renters. As a result of these increasing housing prices, some current residents are forced to move because they can no longer afford rent or taxes. This combination of an incoming wealthy population and outgoing poor population can fundamentally change the character of the neighborhood, even leading to loss of community, and also may have implications for other neighborhoods as displaced residents move into neighborhoods that they can afford (Atkinson, 2000; Freeman et al., 2016). There are many reasons to be concerned with these outcomes (e.g., loss of social capital) (Newman et al., 2016), but our touchstone of concern is this: The environmental justice (EJ) literature demonstrates that Census-measured racial and ethnic minority (and often poor) communities have suffered disproportionately from excess environmental harm, and have also usually suffered from lack of environmentally ameliorative elements (Smith et al., 2020; Iyanda & Lu, 2021; Jelks et al., 2021). Consequently, we want these communities to receive benefits of cleaning and greening.

Because of this potential displacement, a neighborhood's residents might be wary about a city government's plans for environmental improvement, and cities themselves might be wary about making environmental improvements for fear of displacing residents, even in neighborhoods that lack access to environmental amenities. Thus, the threat of potential gentrification may keep residents from supporting environmental amenities, and cities from providing them, obviously a suboptimal outcome for communities that face worse environmental conditions and thus more health risk (Curran & Hamilton, 2017; Rigolon et al., 2020).

Residents are increasingly aware of the risk of green gentrification and displacement. A recent trend has been noted in which poor urban residents may encourage a city to improve environmental conditions only up to a point (Curran & Hamilton, 2012; Wolch et al., 2014). The goal is to improve environmental conditions enough that it benefits the current neighborhood residents, but not so much that properties in the community become appealing to wealthier people from outside the community. This trend of just green enough (JGE) activism (and scholarship) has been noted in recent years, demonstrating the perspective of residents to seek some environmental improvement, but perhaps not as much as would be just (Curran & Hamilton, 2017). This can leave policymakers in a bind; they are likely aware of poor environmental conditions in certain communities, but may feel helpless in being able to improve those conditions if they fear that displacement will begin and environmental improvements will be received by those who already have access to a good environment (Bunce, 2017; Walker, 2021).

Different Neighborhoods, Different Outcomes

However, we argue that the issue is almost certainly not this simple. First of all, it has been widely noted that gentrification and displacement may not be nearly as prevalent as some studies and popular culture suggest (Hochstenbach & Van Gent, 2015; Richardson et al., 2019). Much of the extant research on gentrification has taken place in particular settings—usually large, diverse, high-density cities with high average property values and a large number of wealthy regional residents. For example, much of the early research on gentrification was conducted in cities like New York, Chicago, London, and San Francisco (Freeman & Braconi, 2004; Betancur, 2002; Robinson, 1995). Studies of smaller cities with less diverse populations suggest that, while gentrification can and does occur, it may not be as prevalent in other settings as it is in those special cities (e.g., New York and London are considered world cities) (Richardson et al., 2019; Hwang & Lin, 2016). Secondly, while there has been a connection established between environmental improvement and gentrification, it is clear that not all improvement leads to gentrification, or, more importantly, displacement (Richardson et al., 2019; Eckerd, 2011; Banzhaf & Walsh, 2006). The contextual settings matter; in some regions, in some communities, gentrification and displacement can happen rather quickly, while in other areas green gentrification may take a long time to happen, if it does at all. Third, and related to this last point, cities are constantly in flux, with some amount of residential churn and change in property values expected at all times (Coulton et al., 2012; DeLuca et al., 2019). There is a fine line between community change—for example, due to generational turnover—and residential displacement, and displacement is notoriously difficult to observe (Helbrecht, 2018; Eckerd et al., 2019). But these nuances are often lost in political debate about green gentrification.

In short, we do not know when and under what circumstances green gentrification is likely to occur, and, if it does occur, if it will lead to displacement. If the political debate is focused on the potential for green gentrification and displacement rather than on the need for more equitable access to green amenities, then cities may be exacerbating current inequities in order to avoid something that may not happen in the first place. In previous chapters, Chap. 6 in particular, we explored nuances about when and how green gentrification may occur. Our focus in this chapter is to explore, via a simulation, a set of different contextual circumstances and the resultant potential for environmental gentrification and displacement to occur.

The Model

Our methodological approach in this chapter is to utilize an agent-based model (ABM) to explore a set of different contexts and policy choices and explore the circumstances under which green gentrification displacement seems to occur and occur more quickly. As we described in Chap. 3, we use an ABM because the ABM allows us to alter these contexts. Also, because ABM allows repeated interactions and the observation of emergence, it is one appropriate method for understanding complex adaptive systems (CAS) such as cities (Berry et al., 2002). It also allows us to perform virtual experiments to explore different ways to approach policy options for improving environmental conditions in a community, as we discuss below.

The ABM used in this chapter is built on the foundation of the model described in Chap. 5. All the behavior described in Chap. 5 is the same here; residents seek a location on which to live, using the same utility function with the same similarity preference and budget constraints. Job locations follow the same process, and the real estate market functions the same way. Please refer to the previous chapter for a full review of the basic functioning of the model. In this chapter we will discuss differences from the model as described in Chap. 5.

Neighborhood Gentrification

As with the model described in Chap. 5, a neighborhood can only gentrify if it first becomes 'gentrifiable,' which here we define as a block that has an average wealth of residents in the lowest quartile (this can occur at any point during the simulation run). Defining a neighborhood as gentrifiable basically ensures that any block that is relatively poor could be, but will not necessarily be, gentrified. As a result of our testing of the model in this chapter, we slightly revised how we determine if a neighborhood becomes gentrified. Our definition of gentrification is not identical to, but is in keeping with the literature. Here we assume that any gentrifiable block that subsequently sees its average wealth increase to 10% above the median wealth of all plots is then gentrified (Cf. Richardson et al., 2019), a slight change from Chap. 5, in which we defined gentrification as having occurred when wealth reached the median. This change was practical; in our testing of the model for this chapter, we noted that this threshold was more reflective of empirical evidence regarding the amount of gentrification to expect. Thus, plots can exist in four states: gentrifiable (located in the lowest wealth quartile); gentrified (wealth 10% above average); gentrifying (no longer in the lowest quartile, but not yet gentrified); or none of these (the plot has not become gentrifiable). We use this terminology carefully in describing the model behavior.

The Policy Intervention

The policy intervention in this chapter is the addition of environmental amenities to neighborhoods, based on a set of different scenarios. In Chap. 5, we removed disamenities; we still remove disamenities here, but we also add new, separate, amenities. The idea is that a new amenity placed in a neighborhood will have an independent effect of improving the environmental quality of plots, thus making the plots more desirable according to the residents' utility functions. In the model, then, a combination of low price (owing to previous lack of amenities) and improved environmental quality (owing to the placement of the new amenity) should appeal to any resident randomly selected to consider voluntary relocation, as long as the neighborhood fits their density and similarity preferences and their acceptable wealth levels. These preferences mean that not all wealthy residents will relocate to a gentrifiable plot, but a number of them may, depending on the overall plot attributes and their taste.

Once a plot becomes gentrifiable (the prices are in the lowest quartile), we begin to track movement of residents into and out of the block. One of the advantages of using an agent-based model for a study of gentrification is that it allows us to determine whether and how frequently residents are displaced from gentrifying or gentrified areas. As noted above, some residents are priced out of plots—this occurs when the plot they are currently residing on sees its price rise above the resident's budget constraint. When this happens, residents must relocate—and if it happens in a gentrifying or gentrified area, then they are defined and measured as displaced—and they must select from a random selection of plots within their budget constraint that best maximizes their utility. They must relocate in this case, even if the utility of their current plot is greater than the plot that they will have to move to. We only count residents as 'displaced' if they relocate because of this budget constraint issue and if they currently reside in an area that is first gentrifiable and is then going through the process of gentrifying or is gentrified. Trends associated with the gentrification that occurs, and considerations about what happens to people in the gentrifiable areas, are the key things we track as we compare the scenarios described below.

Simulation Timing

Each simulation trial runs for 75 ticks, or the equivalent of 75 years, although we present the data below showing the period between tick 25 and tick 60. For the first 25 ticks of each trial, we engineer the system to create a city with environmental disparities, such that the poor and especially poor minority residents live under worse environmental conditions than the wealthy and particularly wealthy majority residents. We begin all policy trails after this setup period and for the next 50 ticks. We don't display results after tick 60 because the trends largely stabilized after tick 60.

We are thus considering this as a long-term examination of an urban area, roughly encompassing the life span of one generation, although residents are not static and are constantly entering and exiting the system as they would in a real city. For those who are interested in technical details of the ABM, the Appendix to this chapter provides more information regarding paramaterizations and functions that drive model behavior, most of which are the same as those provided in Chap. 5.

Policy Approaches

Policy Choices

We assess three key policy choices, each drawn from the literature on green gentrification. Previous research has suggested that the approach a city takes to environmental improvement can affect the extent to which gentrification might occur. Since our ABM is intended to be a simple version of a city, we conceive of environmental improvement rather simply as well—the model was designed with "green amenities," conceived of loosely as parks. This choice does fit well with much of the green gentrification literature (e.g., Gould & Lewis 2017) and also fits well with the empirical work in Chap. 6. We then contrast three different policy choices about park placement: the size of parks, the overall desirability of them, and the choice of where to place new parks. Given that we are working with an idealized model, we limit the choices to ideal types so that we can discern the trends associated with each different policy recipe.

Park Size

We look at park size choices as an overall approach to city parks. In the model, cities can opt to locate a small number of large parks or a large number of smaller parks. Our assumption is that a city opting for fewer larger parks will locate parks less frequently than a city opting to locate more, smaller parks. We draw the sizes of parks from a random distribution. A large park policy draws park sizes from a random distribution centered at 6 plots, with a standard deviation of 1. A small park policy draws park sizes from a random distribution centered at 2, also with a standard deviation of 1. Under a small park approach, a new park is placed somewhere at each tick forward (or each year). Under a large park approach, parks are placed once per 5 ticks, (or a new park once every 5 years). According to information provided by the Trust for Public Land, these park-placement frequencies are not unreasonable.

Park Desirability

By park desirability, we refer to the amenities that extend beyond the environmental improvement that parks provide. That is, while larger parks offer more environmental improvement than smaller parks just because of their size, all parks of the same size are conceived of as offering the same amount of environmental improvement to their immediate vicinity. The desirability choice refers to the additional amenities that a park may offer beyond the environmental improvement, such as recreational facilities, extensive playgrounds, cultural events and opportunities, and such. Our idea here, which seems to match well with some of the JGE work, is that the environmental improvement may not actually be the part of the amenity that is most appealing to incomers, and cities may be able to provide the environmental amenity with less concern about displacement if gentrification appeal is more about the other amenities rather than just the environmental improvement.[1] In the model, cities can choose to locate parks that tend to provide a lot of these additional amenities or relatively few of them. We use the same type of variable we described earlier with respect to employment amenities and disamenities. In this case, parks receive a score of 1 if high-desirability parks are chosen or 4 if lower desirability parks are chosen (all parks are amenities, so only one-half of the 1–10 scale described above is used). Parks then influence the environmental quality of nearby areas in the same way as other amenities do.

Park Location

Our last policy choice relates to the approach a city takes in determining where parks should be located. Many cities have zoning requirements for where parks must be located (Mincey et al., 2013), while others may take a piecemeal approach to park location. Whatever the case, our interest with this variable is in exploring how these different choices affect who has access to green amenities and the implications of this. In the model, cities can choose to locate parks according to three different criteria:

- First, parks can be located on the plot of land that is the lowest priced. Such a decision criterion makes sense given that cities often operate under tight park budgets. The likelihood in this case is that parks would be located near where the poorest residents of the city live. In this case, when a new park is created, it will

[1] Gould and Lewis (2017) discuss a case in which a river is cleaned, but the community specifically eschewed park-like walking areas nearby in hopes of gaining environmental benefit without attracting gentrifiers.

select the lowest-priced, currently unoccupied, appropriately sized plot of land upon which to locate at the time.

- Second, parks can be located in the areas with the highest levels of pollution (i.e., lowest plot quality). The assumption in this case is that parks would be located in the areas most in need of environmental improvement. In this case, when a new park is created, it will find the unoccupied set of plots with the highest level of pollution, which will tend to be located in close proximity to environmental disamenities—perhaps not equalizing the environmental harm caused by the disamenity but mitigating it.
- Third, cities can choose to locate parks in "park deserts." or areas that are located far from other parks. In this case, we assume that a city might want to locate parks in areas that do not currently have access to them. Under this scenario in the model, the city locates the park in an area that is currently located the furthest from any other park space.

Given the points that we lay out in Chaps. 1, 2, 3, and 4, our focus here is on the extent of gentrification that takes place, the extent of displacement, and the timing of these. The context of the city is that it is in flux, constantly undergoing change with neighborhoods that are not expected to be static, but can be stable, and our ABM incorporates these realities, with residential movement occurring at every tick. But our analyses focus on the extent to which gentrification-based change in neighborhoods is either rapid or slower-paced, and identify which groups seems to benefit from changing environmental circumstances, and for how long.

We present our results below mostly via line charts comparing outcomes of the different trials that we run. We ran a total of 12 distinct scenarios, with each combination of the two park size policies, the two park desirability policies, and the three location choice options. In Table 7.1, we list each of the scenarios and the codes for the scenarios that we use in the figures below. For ease of visualizing the results we pay particular attention to the first four scenarios listed in Table 7.1.

We focus on the first four scenarios (in bold and color in Table 7.1) as good examples of the most varying of the policy options. Scenario 1 (Small Parks-Lowest

Table 7.1 Scenarios simulated

Scenario 1	**S-LP-LD**	**Scenario 2**	**L-NP-HD**
Scenario 3	**S-NP-HD**	**Scenario 4**	**L-LP-HD**
Scenario 5	L-LP-HD	Scenario 6	L-NP-LD
Scenario 7	L-PD-HD	Scenario 8	S-PD-LD
Scenario 9	S-PD-HD	Scenario 10	L-PD-LD
Scenario 11	S-NP-LD	Scenario 12	L-LP-LD

Key:
First letter: S – small parks scenario; L – large parks scenario
Second set of letters: LP – lowest price plot chosen; NP – Near Pollution (most polluted plot chosen
PD – Park desert chosen
Third set of letters: LD – low desirability parks placed; HD – high desirability parks placed

Price-Low Desirability) is the most straightforward implementation of a JGE approach, as it places many small parks of relatively low desirability in the poorest areas (the JGE literature focuses more on lower-income households than on race/ethnicity). This scenario can be seen as improving the environment without substantially adding recreational or aesthetic amenities (as discussed in footnote 1). Scenario 4 (Large Parks-Lowest Price-High Desirability) presents a contrast to Scenario 1, with large, highly desirable parks placed in the poorest areas; thus it provides an analytically interesting foil to Scenario 1—while still focusing on poor residents, it is not compatible with JGE. Scenarios 2 and 3 are focused on providing the greatest environmental and desirability improvements, but just varying the size of the parks that are built; these may be viewed as something of a reparations approach in which those suffering from the most environmental injustice are being compensated through environmental amenities. The focus on environmental improvement implements our concern with EJ and whether cleaning and greening will swiftly displace residents who have suffered from high levels of pollution. In the figures below, we highlight these four scenarios, but also provide the results for the other scenarios in the background.

We ran each scenario for 50 trials (of 75 ticks each), and calculated the mean scores of each of the key variables within each scenario. Thus, while we could observe wide fluctuation in any set of individual trials, by averaging a large number of runs within each scenario, we smooth out extreme results and compare what *tends* to happen across the different scenarios. The figures below compare the results for each of these 12 scenario recipes.[2]

Results

The results are presented below, focusing on some specific elements of interest. Our first set of comparisons is at the neighborhood level, considering which blocks gentrify and how quickly the process takes place. In our second set of comparisons, we look at the resident level, specifically focusing on who benefits from the environmental quality of the parks that are placed, and for how long, and how frequently minority residents, and the minority poor in particular, are displaced from gentrified areas. Throughout the figures below, we focus on outcomes for minority residents, mostly as a reflection of the fact that gentrification is usually portrayed as displacing

[2] We do also note that we did not run any traditional statistical analysis; this is a choice we prefer in our ABM work. Our rationale is that most inferential statistical techniques like regression assume both linearity of effects and independence of the explanatory variables—though logit and probit techniques do not assume either independence or linearity since all variables are interacted with each other (but analysts frequently seek to interpret them as if they do). In an agent-based model, we assume that effects are interdependent, and non-linear, thus violating key assumptions of the techniques. Beyond this, since all the data we are utilizing are a product of this model, we effectively have a census of our data rather than a sample.

and disrupting poor, minority communities primarily, and also because of our concern with race/ethnicity-based environmental injustice in particular. Since we know that minorities suffer most from environmental injustice, we are most concerned that they not be displaced before they gain benefits from any cleaning and greening. That said, the trends for both majority and minority residents are similar.

Neighborhood Gentrification and Residential Displacement Are Fairly Common in This Simulation

At the outset it is worth noting that gentrification and displacement were both rather likely to occur in all of our scenarios. We see this as a reasonable expectation of a geographically constrained housing market that we observe over a long period of time in a growing city; the churn of the US housing market is well known at this point, and we would expect a housing market characterized neither by complete stability nor by constant instability. It is also worth noting that we did not investigate gentrification solely as an urban-core phenomenon as it is often depicted in the literature. In our model, gentrification could occur anywhere in the region, not just in the denser urban core neighborhoods—thus we effectively enabled suburban gentrification as well as urban gentrification. This is well since gentrification of at least inner suburbs is increasingly visible and of concern to urbanists (Rankin & McLean, 2015; Markley, 2018).

With these caveats noted, we do see some variation in the amount of gentrification that occurs across the different scenarios. Figure 7.1 illustrates the proportion of the plots that become gentrified in each scenario at any point during the simulation, and we see that all of the scenarios are clustered between about 35% and 45% of all land being gentrified by the end the trial runs. There are not any very strong trends with respect to which scenarios tend to lead to more or less gentrification, although there may be a slight tendency for a bit more gentrification to occur in the scenarios where smaller parks are sited rather than larger parks, particularly for our four featured scenarios, and the pace of gentrification (the slope of the lines) is also similar across scenarios. On the other hand, the blue L-NP-HD line consistently shows more gentrification than the green L-LP-HD, which performs better than all other featured scenarios after about tick 30. Considering all of these results together supports the idea that we can't look at one variable at a time but must look at configurations.

The differences between scenarios become a bit more pronounced if we look just at the proportion of minorities that are located in gentrified areas, as shown in Fig. 7.2. The results are still clustered together, but the range of minorities in gentrified areas is larger (about 25%–40%) and between ticks 30 and about 38 we see a slight trend of more minorities living in gentrified areas when smaller parks are sited than when larger parks are sited; in these ticks at first the yellow S-NP-HD and red S-LP-LD lines exhibit more minorities in gentrified areas than does the blue

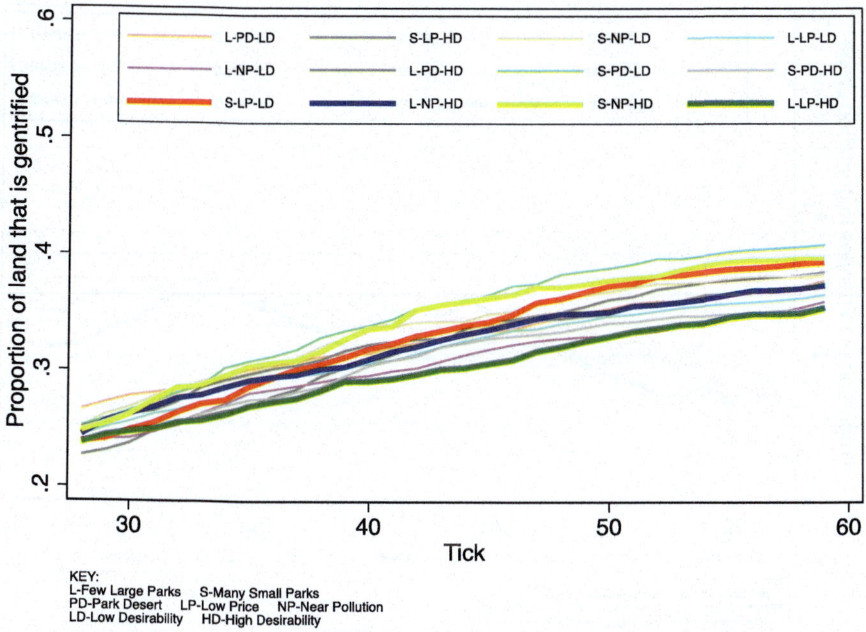

Fig. 7.1 Proportion of land that is gentrified across scenarios

L-NP-HD line. By tick 40, however, these trends no longer hold. These results emphasize that timing also matters. Gaining 8 years in a cleaner, wealthier area could be of benefit to polluted minority communities.

After about tick 45, the general trends seem to be mostly flat (there is perturbation, but no trend of growth or shrinkage), which is to say that scenarios reach a point of overall stability with respect to the proportion of minorities that are living in gentrified areas at any given time.

If we look at class rather than race, we again see the results clustered together. Figure 7.3 shows the proportion of all poor residents (of both "races") living in gentrified areas, and there are few discernable trends among our four featured scenarios as their lines cross and re-cross each other. Although we do not highlight the results when park desert locations are chosen, the park desert scenarios seem to result in slightly more poor residents being located in gentrified areas (e.g., the orange line hovers near the top), but this trend is not very pronounced. Comparing Figs. 7.2 and 7.3 demonstrates that there are more discernable trends for minority residents than if we consider all poor residents together.

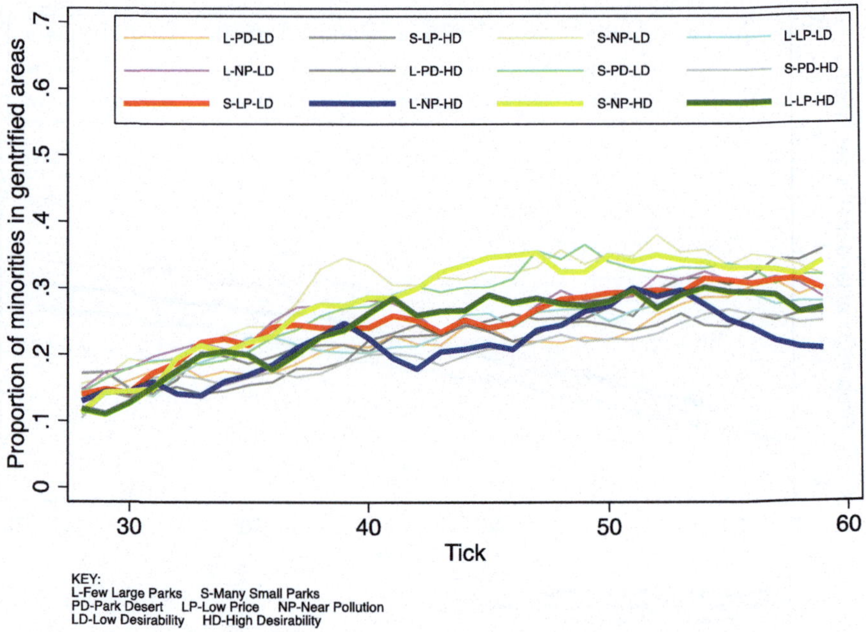

Fig. 7.2 Proportion of minorities in gentrified areas

Displacement

Shifting the outcome variable to focus on displacement, we start to see more variation between scenario recipes. Figure 7.4 presents the proportion of minority residents that are displaced at least once at any time during the simulation, and we see more pronounced trends. Figures 7.1, 7.2, and 7.3 represent the picture of gentrification in process. Figure 7.4 presents the specific outcome of concern: displacement of residents. Once again, at least after the first few years, we tend to see that more minority residents are displaced at least once when more, smaller parks are sited than when fewer, larger parks are (although this is not entirely consistent across all scenarios). Up through around tick 48, the red S-LP-LD scenario, which we see as the most similar to JGE, performs worst, though the yellow S-NP-HD, which might be considered a reparations scenario, performs worst after tick 50. The two scenarios with the most common experience of displacement involve small parks, and the scenario with the highest experience of displacement is the one in which the most environmental improvement takes place (targeting polluted areas with high desirability parks) and where smaller parks are chosen. However, it is notable that in all of these cases, at least for 10 years, the proportion of minorities displaced at least once is low, including for L-NP-HD, which suggests that at least sometimes cleaning and greening is also reasonably sustainable for local residents. Also, the scenarios suggest that placing high desirability parks in park-desert areas generates the least minority displacement of all, another hopeful finding (see the lighter-gray and lavender-gray lines running below the dashed Average line).

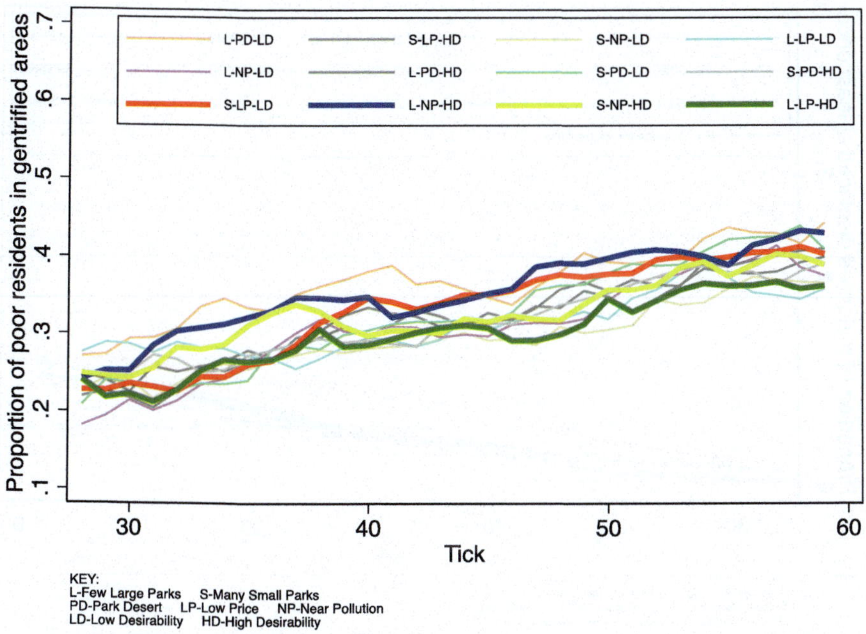

Fig. 7.3 Proportion of poor residents in gentrified areas

In Figs. 7.4 and 7.5 we have added an additional line, the black dashed line, which provides the average for all residents (minority plus majority) across all scenarios. In Fig. 7.4 this line indicates that the proportion of those displaced at least once is low on average for the first 10 years.

Displaced Residents Tend to Experience Displacement Repeatedly

Figure 7.5 looks at those who have been displaced more than once, examining how many times minority residents experience displacement. The lines in Fig. 7.5 show the average number of times a minority resident was displaced, if they were displaced at least once—the intuition here is that displacement may not be a one-time experience. If someone is displaced, they may end up moving to another area at risk of gentrification, and thus may end up getting displaced a second time (or more); this matches the experience of Mitchell (2019), discussed in Chap. 1, who states that she and her husband were displaced 4 times in 5 years.

There are two things to note immediately in Fig. 7.5. First, we see that displacement is a common occurrence for those minority residents displaced at least once. If a minority resident is displaced at least once, then they were likely displaced at least a second, and perhaps a third time over the entire course of the model's time period. Figure 7.5 is also the most definitive result, as we see two consistent clusters of scenarios appear—when small parks are sited frequently, minority residents tend

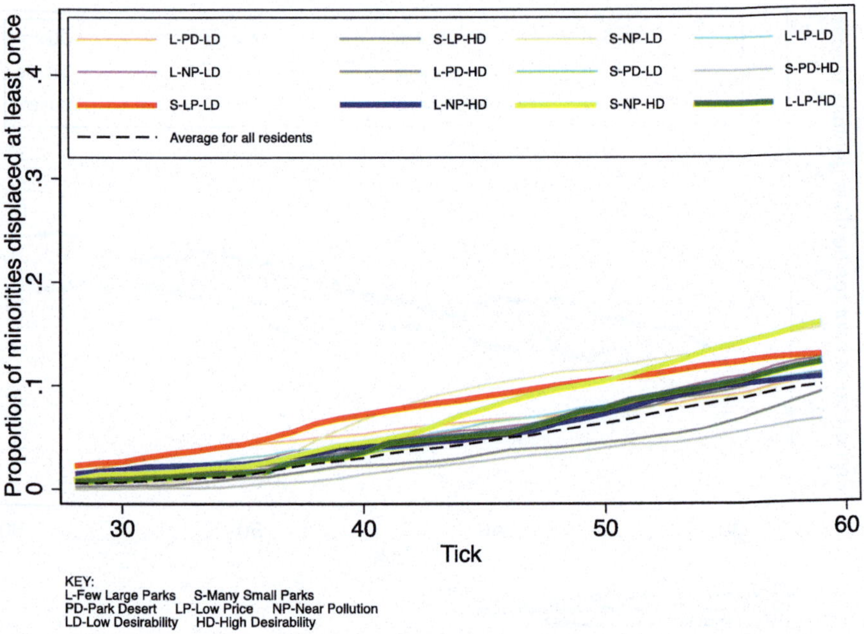

Fig. 7.4 Proportion of minorities displaced at least once

to be displaced more times than when large parks are sited less frequently. Each of the lines approaching 3 times displaced is a small park scenario and each of the lines closer to 2 is a large park scenario. The black dashed line is the average number of times all residents (majority plus minority) experience displacement across all scenarios, so these small park scenarios lead minority residents to experience displacement more than the average resident.

Putting these numbers in context with results from Fig. 7.4, we see that the *small, near pollution, high desirability* (yellow) scenario results in the highest proportion of minorities being displaced (about 12%), with those who are displaced being displaced more often (about 3 times total). In comparison, the *large, near pollution, high desirability* (blue) scenario results in fewer displacements (less than 10%) and fewer times displaced for that group (about 2 times). Perhaps paradoxically, In Fig. 7.5 we find that a JGE scenario (red), in which small, low desirability parks are sited in poor areas, eventually results in the more frequent experience of minority displacement.

Displacement Takes Place at Different Time Intervals

Figure 7.6 approaches a similar idea, but in a different way. The lines in Fig. 7.6 indicate the tenure of residents in gentrifying neighborhoods, or in other words, the number of ticks that it took for a resident to be displaced—that is, how long did the process

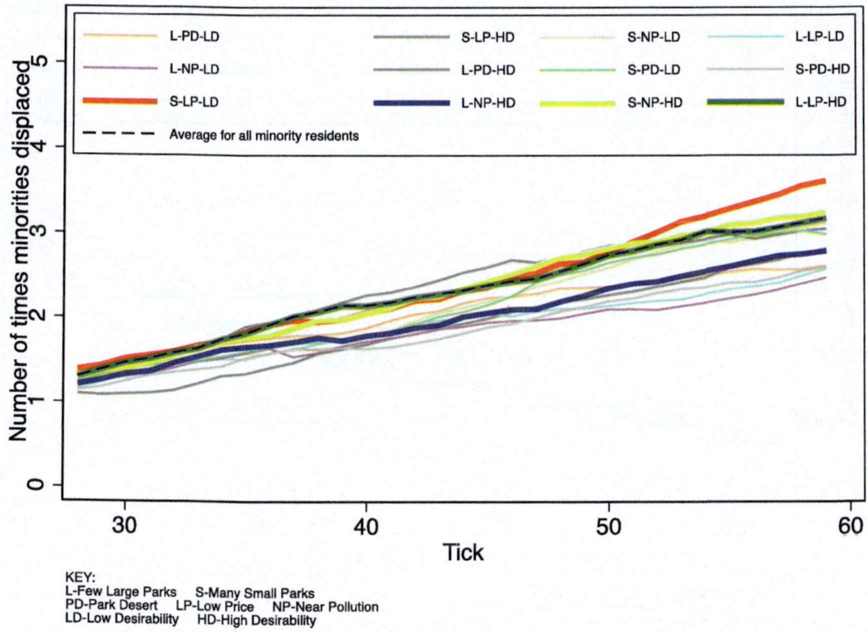

Fig. 7.5 Average number of times minorities are displaced—if displaced at least once

of being displaced take once the neighborhood began to gentrify (or the point at which a gentrifiable plot saw enough of a value increase to move above the first quartile of wealth). Once a gentrifiable plot's price rose above the first quartile's upper bound, we began measuring the tenure of residents on their home plot. Then, once the plot gentrified, we assessed how long until residents were displaced. Basically, this is a measure of how long the resident is able to access the benefits of a healthier and wealthier community that is increasing in land value before they are displaced.

Figure 7.6 is an interesting contrast to Fig. 7.5. While Fig. 7.5 tended to show that smaller parks may result in more frequent displacement, Fig. 7.6 suggests that this frequency should be contextualized. The JGE scenario (S-LP-LD) may generally see more displacement, but that displacement takes place over a longer period of time. This suggests that smaller, less desirable parks foster more frequent displacement for those displaced, but at the same time offer residents who will be displaced longer tenure in the neighborhood before they are displaced. When parks are sited in the lowest-priced plots, the neighborhoods that parks are cited in tend to be the poorest in the simulation. This suggests that gentrification may take longer to play out in the poorest of neighborhoods. Tenure tends to be lower in all three featured scenarios when the park selection choice is the most polluted (blue, green, yellow). The most polluted land may not be particularly low-priced land. Considered by itself, we would expect polluted land to be low-priced. But, because residential demand importantly drives price changes in land, and some highly polluted areas may be relatively remote from existing residential development, polluted land may

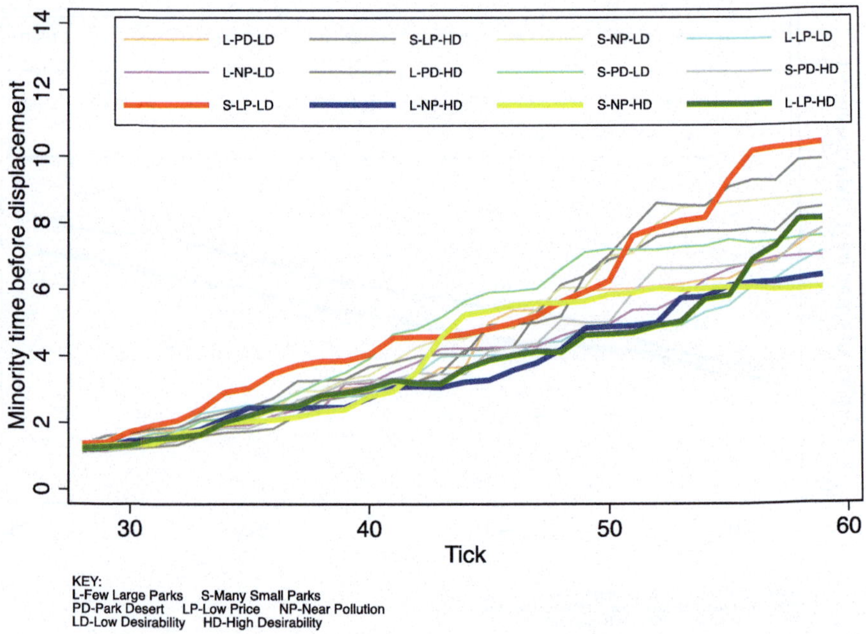

Fig. 7.6 Tenure from start of gentrification process until displacement for minority residents

not necessarily exhibit the lowest prices in the region since there may not be much residential development nearby.

We find a bit of a paradox with the *small, lowest price, low desirability* scenario, which is the scenario most consistent with the JGE idea. Under this scenario, nearly 20% of all minorities are displaced at least once, and those who get displaced once are likely to be displaced again. However, while this scenario also results in the most times that minority residents get displaced, it also allows them the longest time in a gentrifying neighborhood before that displacement happens. In effect, there is a much more gradual process of gentrification in these cases, so the JGE strategy may be effective in slowing down the gentrification process. At the same time, however, displacement is rather common in this scenario and, when it happens, it happens to residents more times than in the other scenarios. These results do stress the importance of understanding not just the beginning and end states of gentrification and displacement, but also how these processes play out over time.

Displacement, Though Clearly a Burden, Might Have Some Benefits

Finally, in Figs. 7.7 and 7.8, we consider what the end result of displacement is for those minority residents who were displaced at least once. In Fig. 7.7, we consider the change of the environmental quality that displaced residents experienced after being displaced, with environmental quality, as described above, an attribute of the

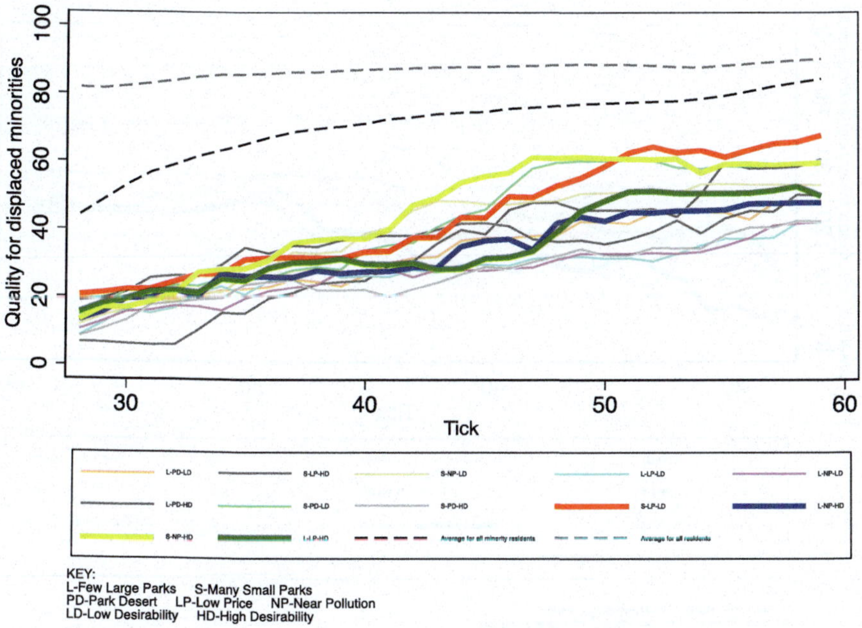

Fig. 7.7 Quality for displaced minorities

plot the resident is located on with a relative score between 0 and 100 (with higher values indicating better environmental quality). The scenario lines in Fig. 7.7 track the total average quality level of minority residents after they have been displaced. In Fig. 7.8, we track how their wealth changed after they were displaced. These figures allow us to see if the displaced faced an increased environmental burden when they were forced to move, and secondly to consider if the displacement allowed them to build any wealth (assuming the sale of property when they were displaced).

The changes in quality over time shown in Fig. 7.7 track consistently with the overall change of quality as the model moves forward. The dashed gray line added to the top of the graph is the overall average level of quality for all residents, and the dashed black line is the overall average level of quality for all minority residents in all scenarios. First, note that minority quality levels overall (shown by the dashed black line) start out significantly below those of the average over all residents (shown by the higher, gray, dashed line); this is the EJ effect. Also the quality levels for displaced minorities start out significantly below the average quality levels for all minorities. The simulations are engineered to see environmental improvement as the model ticks forward in time and parks are added (and disamenities cleaned up), so the upward trajectory is expected. We can see that while the trend of quality improvement for displaced minority residents (shown by the not-dashed lines) is similar to the overall trend of quality improvement, the quality variable is lower than average at the start of the simulation, and the gap is never closed with respect

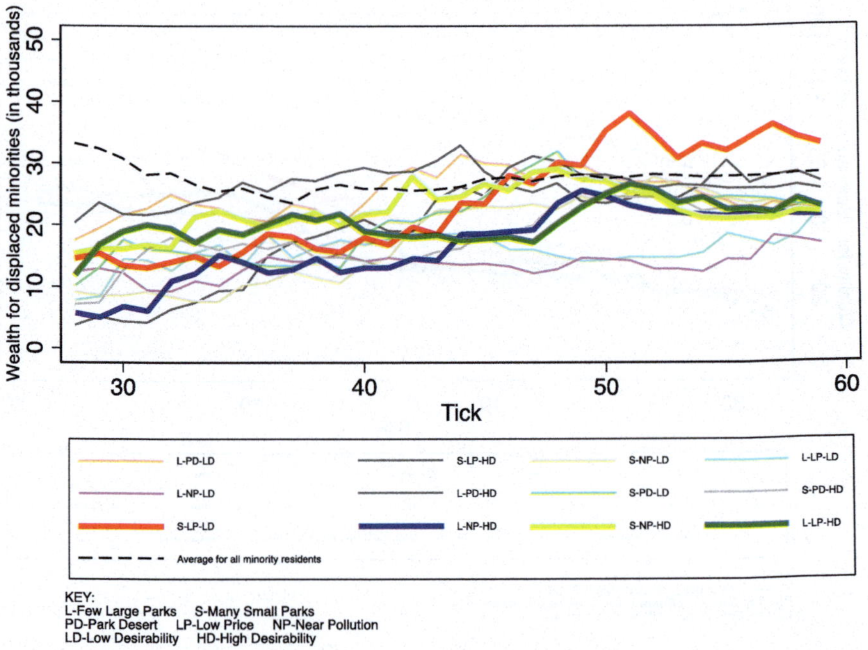

Fig. 7.8 Wealth for displaced minorities

to displaced residents (the gap is closing with respect to the full minority population). However, some scenarios do better: the *small, lowest price, low desirability* scenario is estimated eventually to gain slightly relative to the overall average (and in some time periods, tick 37 to tick 50, S-NP-HD does best).

A difference is much more discernable with the wealth variable. Here, the dashed black line has been added as a comparison to the overall average wealth of all minority residents as the simulation moves forward. We would expect the black line to be relatively flat, as houses are sold and some gain wealth and some lose, and indeed the black line is mostly flat after tick 40, with just a slight upward trajectory. The displaced all start out with wealth well below the average level, but in some cases the displaced show wealth gains relative to the overall minority population. The displaced minority residents under the JGE scenario actually end up with more wealth than average. In fact, in every scenario the displaced end up at least marginally better off in terms of wealth during the trials, while the average resident ends up with about the same or slightly less wealth, suggesting that, in the aggregate, environmental improvement, if widespread, may well result in displaced minority residents reaping more benefits relative to other residents.[3]

[3] But please do keep in mind that this set of scenarios treats all residents as owners. See Chap. 8 for comparison.

Discussion

In our artificial world, we find that gentrification is prevalent regardless of the policy scenario. This is likely due in part to how we define gentrification, with blocks that see a wealth change from the first quartile of resident wealth to 10% above the median being what we consider gentrified. A more restrictive approach, say a move from the first to the fourth quartile, might have resulted in less overall gentrification, but we opted for the more inclusive definition to ensure that we would have a considerable amount of gentrification to observe (also, this choice reasonably matches other's definitions, such as that used by Richardson et al., 2019).

Many of the scenarios were also quite similar with respect to the gentrification trends overall. The overall trends of gentrification and displacement tended to play out similarly regardless of how the city implemented policies to improve environmental conditions. Again, this is not really surprising as we found in previous work that density preferences, in particular, were influential in gentrification occurring—i.e., when people have higher density preferences, gentrification is more likely—and, though individual residents have varying density preferences, we did not vary the distribution of density preferences in these trials. That said, there were some differences, particularly with respect to the timing of the process. As we noted in Chaps. 3 and 4, we are interested in how different approaches might enable gentrification to proceed more slowly, allowing residents to benefit from improved environments, and also giving residents the time to plan for the coming changes, perhaps lessening the burden on them when they are displaced. Our findings offer some rather clear distinctions in the scenarios where gentrification occurs more quickly or more slowly, although not without some caveats.

Our most interesting results relate to the JGE scenario. Ultimately, we find that the results for a JGE policy strategy are somewhat mixed. More minority residents experience displacement, and those residents that are displaced tend to be displaced multiple times when this strategy is used. At the same time, however, most minority residents in this scenario are displaced at a slower rate, with the process potentially taking years to play out, and when they are displaced they might actually experience gains in both environmental quality and wealth. A caveat to this last point is that the model used in this chapter only included residents who owned their homes (i.e., no renters), so the experience of displacement at least tended to increase their wealth. In Chap. 8 we consider different effects for owners and renters. We also note that the model does not include an accounting of the burden of displacement (search costs, moving costs, loss of community), so we want to be careful proclaiming this as a good thing, but what we see is that the result of a JGE strategy may be more nuanced than it often appears to be.

Conclusion and Policy Implications

Yet, we want to be careful with these results, because the policy choices they might imply need to be thought out. The first implication is that either providing no environmental improvement or minimal environmental improvement might improve community stability. This is not a helpful interpretation in our view, because the potential health benefits of the improvement need to be considered along with the stability issue. Environmentally hazardous conditions create problems that we see as greater than those associated with displacement risk. Both are problematic, of course, but poor environmental conditions can affect the health (and mortality) of residents, the ability of children to learn and do well in school, and a poor environment has also been associated with other neighborhood stressors such as crime and infrastructure disinvestment. We want to be clear that the goal of thwarting gentrification is not ethically equal to the goal of ensuring the health, safety, and welfare of residents. These are both worthy goals, but we do not see them as equivalent.

With this caveat in mind, we do think it is a worthy policy goal to both reduce the likelihood and increase the duration of the gentrification process and, thus, delay displacement. Our model implies some good approaches to doing so. First off, in the real world, at least over time, cities will utilize a mixed approach rather than one of our more rigid scenarios, so the question here is what to emphasize in policy, given that reality. Although we refer to them as parks in our model, these land uses could be thought of as any combination of environmental improvement with development. In the model, we differentiated between large, highly desirable amenities, and small, less desirable amenities. The parks did not vary in the extent to which they improved environmental conditions (although of course larger parks would have a geographically broader effect), but they did vary in the extent to which they were desirable. We conceived of this as the difference between a simple community park that might include basic park spaces like playgrounds and walking spaces, and those with more amenities like music venues or cafes. Our results suggest that the impact of just the environmental improvement without attendant amenities might ease the gentrification process. It may be that large recreational amenities have more of a profound effect on gentrification than the environmental improvement. Although, it is worth noting that environmentally focused scenarios do not work poorly across the board: There are some scenarios where the context and unique configurations of settings enable environmental improvement that does not lead to extensive displacement.

Overall then, we see this as relatively straightforward support for the value of a JGE approach—provided that 'green enough' means that all the environmental quality improvement possible is indeed produced, and it is "shiny objects" that are left off. Thus, the tradeoff is not with environmental improvement but is rather a tradeoff of other choices made about the amenities. This is not to suggest that parks in poor communities should be less prevalent or of lesser health value, but rather that they include basic park features but perhaps not additional amenities that are beyond this basic setup and more attractive to in-comers.

We do note that this is a suboptimal recommendation. We are, in effect, arguing that communities that are at risk of gentrification be given fewer amenities than wealthier areas, which is an uncomfortable recommendation to make. This is anti-Rawlsian as it places the burden of the tradeoff on the least well-off set of people. Yet, burden is unfortunately unavoidable, and so the question is which tradeoff minimizes the burden. Our rationale, then, is straightforward—we see reducing park amenities and park sizes as the least burdensome of the other choices, most notably the choices of either living with poor environmental conditions or being displaced quickly due to gentrification. A JGE approach will not stop gentrification, but it will make the process more gradual, which we see as the most feasible approach to managing what occurs in a complex adaptive system.

Appendix: ABM Details[4]

Utility Function

All residents, including those present at the initialization of the model, select a location starting with a basic utility function with price, quality, and distance as parameters. The base utility function is shown in Eq. (7.A1):

$$u_{i,t} = p_{i,t}^{-\alpha} \cdot q_{i,t}^{\beta} \cdot d_{ik,t}^{-\gamma} \tag{7.A1}$$

The overall utility rate, u, for plot i at time t is a function of price, quality, and distance to amenities and disamenities. Residents seek an equal balance of the lowest price, p, of plot i at time t, the highest environmental quality, q, and the shortest distance to some amenity, k. In this case, residents seek to minimize distances to either amenity or disamenity land use, assuming that all land uses offer some benefit to the resident, such as a place to work or shop. The parameters α, β, and γ are weighted equally, assuming that residents equally balance these considerations (Torrens & Nara, 2007).

Resident choices, and this utility function, are constrained by three different factors. First, we assume that residents will not locate in a plot that they cannot afford, nor in one that is substantially lower than they can afford. We operationalize this in the model by assuming that residents will not select a plot that is more than 3 times greater than their individual wealth level, nor a plot that is less than their wealth.

Residents are also constrained by a similarity preference—basically an implementation of Schelling's (1971) insights. In line with Schelling, we assume that residents will limit their choice-set of locations to ensure that some proportion of other residents near them is of the same kind as themselves. If no plots are available that fit this constraint, residents will either remain at their current location, or, if they

[4] Descriptions in the appendix are slightly modified from Eckerd et al. (2019).

are new to the region, they will relax the similarity preference and select a plot that fits their price constraint and maximizes their utility. Finally, residents have a density preference. Each resident draws a density preference from a normal distribution and will constrain their plot choices to locations such that they will share the plot with a number of residents equal to or less than their density preference.

After these constraints have been satisfied, all newly introduced residents move to the plot that has the highest utility score in their available choice set. Existing residents are mobile and may or may not relocate depending on a comparison between their current home and other plots in the region. As the simulation steps forward in time, at each time-tick 10% of the existing residents are randomly selected to consider relocating and all displaced residents must relocate. Those selected seek a location that fits their price, similarity, and density constraints and scores the highest utility value as described above. If they find a location with a higher utility value than their current location, they relocate. If there are no plots in their constrained choice set that increase utility above their current location, they do not move, unless they have been displaced.

Land Price and Quality

Plots of land are modeled to function according to a real estate market, calibrated to function as realistically as possible given the model's parsimonious design. Each plot's price is affected by residential location decisions nearby and the presence of amenities and disamenities. First, plot prices are decreased based on the inverse distance to disamenity land use and increased by distance to amenity land use. Plot prices are also weighted at each tick by the average wealth of the residents located on the plot, and those on the eight contiguous plots (Veldkamp & Verburg, 2004), as shown in Eq. (7.A2):

$$p_{m,t} = \left[p_{m,t-1}\right] + \left[\frac{\Sigma w_{i,t}}{n_i} + \frac{\Sigma w_{j,t}}{n_j}\right] \tag{7.A2}$$

In Eq. (7.A2), the price, p, of plot m at time t is a function of its price at time $t-1$, balanced by the average wealth, w, of the residents, i, located on the plot and the average wealth of the residents, j, located on the eight contiguous plots. Basically, plots with wealthier residents nearby cost more (and vice versa).

Quality is adjusted similarly, although this value is not affected by residents and is only affected directly by the plot's proximity to amenities, including parks, and disamenities. Quality is adjusted up or down based on the resident-proximity preference score of nearby land uses, as shown in Eqs. (7.A3a and 7.A3b):

$$q_{m,t} = \left[q_{m,t-1} + \left(q_{m,t-1} \left(.01 * s_{k,t} \right) \right) \right] + \left[q_{m,t-1} + \frac{\Sigma q_{j,t}}{n_j} \right] \quad (7.A3a)$$

$$q_{m,t} = \left[q_{m,t-1} - \left(q_{m,t-1} \left(.01 * s_{k,t} \right) \right) \right] + \left[q_{m,t-1} - \frac{\Sigma q_{j,t}}{n_j} \right] \quad (7.A3b)$$

In these equations the quality q for plot m at time t is increased as in Eq. (7.A3a), or decreased based on Eq. (7.A3b), depending on if the closest land use, k, is an amenity or disamenity. Quality is then altered by comparing the focal plot's quality level with the average quality of the eight contiguous plots, j, either raising or decreasing quality by the difference between the focal plot's quality and the average quality nearby, diffusing it at a rate of 0.70 (i.e., the surrounding plots are affected by the price and quality of a focal plot, with affected plot's price and quality values being adjusted by 70% of the average of all surrounding plots) (Table 7.A1).

Table 7.A1 Base model paramaterizations

Parameters	Value	References
The simulation world size	10,201 plots	Authors' assumption
Growth rate	5%	Authors' assumption (growing region; e.g., sunbelt)
Initial quality of plots (0–100)	50	Hypothetical value
Initial price of plots: mean, standard deviation (a normal distribution)	\$173,000, \$34,000	US median housing price (US Census Bureau, 2010)
Resident jobs per firm	50:1	Authors' assumption
Initial number of residents	200	Authors' assumption
Initial majority | Minority composition	70% | 30%	Authors' assumption
Base majority income: Mean, standard deviation (a gamma distribution)	\$54,000, \$41,000	US Income, White non-Hispanics (US Census Bureau, 2010)
Base minority income: Mean, standard deviation (a gamma distribution)	\$32,000, \$40,000	US Income, African-Americans (US Census Bureau, 2010)
Plot price and quality diffusion rate	0.7	Veldkamp and Verburg (2004), Brown et al. (2005)
Price and quality exponential decay rate (δ)	1.5	Veldkamp and Verburg (2004)
Utility balancing parameters (α, β, γ)	0.5	Brown and Robinson (2006), Torrens and Nara (2007)
Density preference, standard deviation	10, 3	Authors' assumption
Size of resident choice set when relocating	10	Authors' assumption/model calibration
Resident similarity preferences	25%	Conservative estimate based on Clark (1991)

References

Atkinson, R. (2000). The hidden costs of gentrification: Displacement in Central London. *Journal of Housing and the Built Environment, 15*, 307–326.

Banzhaf, S. H., & Walsh, R. P. (2006). *Do people vote with their feet? An empirical test of environmental gentrification* (NBER, RFF Discussion Paper No. 06-10).

Berry, B. J., Kiel, L. D., & Elliott, E. (2002). Adaptive agents, intelligence, and emergent human organization: Capturing complexity through agent-based modeling. *Proceedings of the National Academy of Sciences, 99*(suppl 3), 7187–7188.

Betancur, J. J. (2002). The politics of gentrification: The case of west town in Chicago. *Urban Affairs Review, 37*(6), 780–814.

Brown, D., & Robinson, D. (2006). Effects of heterogeneity in residential preferences on an agent-based model of urban sprawl. *Ecology and Society, 11*(1), 46.

Brown, D., Page, S., Riolo, R., Zellner, M., & Rand, W. (2005). Path dependence and the validation of agent-based spatial models of land use. *International Journal of Geographical Information Science, 19*, 153–174.

Bunce, S. (2017). *Sustainability policy, planning and gentrification in cities*. Routledge.

Clark, W. A. (1991). Residential preferences and neighborhood racial segregation: A test of the Schelling segregation model. *Demography, 28*, 1–19.

Coulton, C., Theodos, B., & Turner, M. A. (2012). Residential mobility and neighborhood change: Real neighborhoods under the microscope. *City, 14*(3), 55–89.

Curran, W., & Hamilton, T. (2012). Just green enough: Contesting environmental gentrification in Greenpoint, Brooklyn. *Local Environment, 17*(9), 1027–1042.

Curran, W., & Hamilton, T. (Eds.). (2017). *Just green enough: Urban development and environmental gentrification*. Routledge.

DeLuca, S., Wood, H., & Rosenblatt, P. (2019). Why poor families move (and where they go): Reactive mobility and residential decisions. *City & Community, 18*(2), 556–593.

Eckerd, A. (2011). Cleaning up without clearing out? A spatial assessment of environmental gentrification. *Urban Affairs Review, 47*(1), 31–59.

Eckerd, A., Kim, Y., & Campbell, H. (2019). Gentrification and displacement: Modeling a complex urban process. *Housing Policy Debate, 29*(2), 273–295.

Freeman, L., & Braconi, F. (2004). Gentrification and displacement New York City in the 1990s. *Journal of the American Planning Association, 70*(1), 39–52.

Freeman, L., Cassola, A., & Cai, T. (2016). Displacement and gentrification in England and Wales: A quasi-experimental approach. *Urban Studies, 53*(13), 2797–2814.

Gould, K., & Lewis, T. (2017). *Green gentrification: Urban sustainability and the struggle for environmental justice*. Routledge.

Helbrecht, I. (2018). Gentrification and displacement. In *Gentrification and resistance: Researching displacement processes and adaption strategies* (pp. 1–7). Springer.

Hochstenbach, C., & Van Gent, W. P. (2015). An anatomy of gentrification processes: Variegating causes of neighbourhood change. *Environment and Planning A: Economy and Space, 47*(7), 1480–1501.

Hwang, J., & Lin, J. (2016). What have we learned about the causes of recent gentrification? *City, 18*(3), 9–26.

Iyanda, A. E., & Lu, Y. (2021). Perceived impact of gentrification on health and well-being: Exploring social capital and coping strategies in gentrifying neighborhoods. *The Professional Geographer, 73*(4), 713–724.

Jelks, N. T. O., Jennings, V., & Rigolon, A. (2021). Green gentrification and health: A scoping review. *International Journal of Environmental Research and Public Health, 18*(3), 907.

Markley, S. (2018). Suburban gentrification? Examining the geographies of new urbanism in Atlanta's inner suburbs. *Urban Geography, 39*(4), 606–630.

Mincey, S. K., Schmitt-Harsh, M., & Thurau, R. (2013). Zoning, land use, and urban tree canopy cover: The importance of scale. *Urban Forestry & Urban Greening, 12*(2), 191–199.

Mitchell, S. (2019, March 18). *In Richmond, Virginia, gentrification is colonization*. NCRC. https://ncrc.org/gentrification-richmondva/

Newman, B. J., Velez, Y., & Pearson-Merkowitz, S. (2016). Diversity of a different kind: Gentrification and its impact on social capital and political participation in Black communities. *Journal of Race, Ethnicity, and Politics, 1*(2), 316–347.

Rankin, K. N., & McLean, H. (2015). Governing the commercial streets of the city: New terrains of disinvestment and gentrification in Toronto's inner suburbs. *Antipode, 47*(1), 216–239.

Richardson, J., Mitchell, B., & Franco, J. (2019). *Shifting neighborhoods: Gentrification and cultural displacement in American cities*. National Community Reinvestment Coalition Report.

Rigolon, A., Keith, S. J., Harris, B., Mullenbach, L. E., Larson, L. R., & Rushing, J. (2020). More than "just green enough": Helping park professionals achieve equitable greening and limit environmental gentrification. *Journal of Park & Recreation Administration, 38*(3), 29–54.

Robinson, T. (1995). Gentrification and grassroots resistance in San Francisco's tenderloin. *Urban Affairs Quarterly, 30*(4), 483–513.

Schelling, T. C. (1971). Dynamic models of segregation. *Journal of Mathematical Sociology, 1*(2), 143–186.

Smith, G. S., Breakstone, H., Dean, L. T., & Thorpe, R. J. (2020). Impacts of gentrification on health in the US: A systematic review of the literature. *Journal of Urban Health, 97*, 845–856.

Torrens, P., & Nara, A. (2007). Modeling gentrification dynamics: A hybrid approach. *Computers, Environment and Urban Systems, 31*, 337–361.

U.S. Census Bureau. (2010). *2010 Decennial Census. U.S. Department of Commerce*. Retrieved September, 2022, from https://data.census.gov/

Veldkamp, A., & Verburg, P. (2004). Modeling land use change and environmental impact. *Journal of Environmental Management, 72*(1–2), 1–3.

Walker, R. H. (2021). Engineering gentrification: Urban redevelopment, sustainability policy, and green stormwater infrastructure in Minneapolis. *Journal of Environmental Policy & Planning, 23*(5), 646–664.

Wolch, J. R., Byrne, J., & Newell, J. P. (2014). Urban green space, public health, and environmental justice: The challenge of making cities 'just green enough'. *Landscape and Urban Planning, 125*, 234–244.

Chapter 8
Mitigating Displacement from Green Gentrification: Examining the Role of Housing-Cost Control Measures

Soomi Lee and Adam Eckerd

Abstract As discussed in earlier chapters, the most common recommendations for reducing the likelihood of gentrification with displacement involve some type of housing-cost control. Some methods of such control involve preventing or reducing the rapid rise of property taxes and/or rent. And 16 US states, plus Washington, DC, currently have some such type of control, most commonly for homeowners, while more than 30 states forbid at least some controls—disallowing rent controls, for example is common. In this chapter, we review the empirical situation of such rent regulations, tax caps, and circuit-breakers, and then we perform Agent-Based Modeling (ABM) analysis to understand their effects in reducing green gentrification and displacement. The findings indicate two main themes. First, virtually any policy effort, whether using a just green enough (JGE) approach or implementing any combination of property tax caps or rent caps, results in less harmful displacement effects than simply following an environmentally focused green intervention.

Introduction

Previous research has shown empirical evidence that green projects may increase nearby housing values and foster cycles of gentrification and displacement. In the American context, this often means that in-movers are wealthier, whiter, and more educated, while out-movers are poorer and people of color. However, well-known data and analytical challenges exist to cloud our judgment as to whether green projects cause gentrification and displacement of current residents (Zuk et al., 2015). A review paper indicates that environmental improvement could lead to what is termed

S. Lee
University of La Verne, La Verne, CA, USA
e-mail: slee4@laverne.edu

A. Eckerd (✉)
Old Dominion University, Norfolk, VA, USA
e-mail: aeckerd@odu.edu

H. E. Campbell et al., *Green Gentrification and Environmental Injustice*,
https://doi.org/10.1007/978-3-031-65100-7_8

"exclusionary displacement" and potentially displace some renters (Zuk et al., 2015), although the extent of displacement likely differs significantly across neighborhoods (Newman & Wiley, 2006).

In previous chapters, we have explored issues of environmental injustice relating to environmental improvement, and we note that gentrification that plays out over a longer process may lessen the harm that gentrification-related displacement causes. In this chapter, we present that, in the US, many cities and states already have policy instruments to reduce displacement. However, the policy environment is heterogeneous, potentially affecting the distribution of displacement risks for subpopulations. As such, it is important to understand the effect of policies that improve environmental conditions (e.g., greening and cleaning) when they are implemented into policy environments that already contain housing protection policies. For instance, some cities (e.g., New York City and Los Angeles) have laws to protect homeowners and renters from displacement, and others do not. However, the green gentrification literature has generally ignored this reality, including our earlier chapters in this work.

This chapter extends previous research by examining policy environments with pre-existing policies that intend to keep residents from being displaced. One crucial difference this entails is acknowledging the difference between renters and homeowners, as existing housing policies are predicated on this distinction. Existing studies allude to the distinction, but not explicitly in conjunction with policies designed to protect residents. In addition, some prior studies have recommended tenant (renter) protection laws, land trusts, housing cooperatives, and social housing construction (e.g., Jelks et al., 2021; Anguelovski et al., 2019; Anguelovski et al., 2022), although the degrees to which these policies actually protect current residents has been difficult to document (Quinton et al., 2022). However, laws that can protect homeowners in gentrifying areas have yet to be sufficiently examined in the green gentrification literature. Additionally, policy recommendations for tenant protection have often been made without considering differing and complex state and local policy environments.

To fill the gap, we reexamine how the current literature discusses green gentrification and displacement in the context of the local policy environment, present protective policy measures, especially property tax laws and rent controls, to stabilize neighborhoods, and use agent-based modeling (ABM) to track the residential mobility of homeowners and renters in various policy environment.

Policies to Limit Displacement

New York City offers a good example of policies that are intended to limit displacement for both renters and owners and how they function in an environmental gentrification context. Black and Richards (2020) note that increased property values that resulted from the establishment of New York City's High Line Park project demonstrate that green space access is often only affordable to the wealthy, which typically leaves the poor bearing the weight of environmental disamenities. However, the paper does not directly examine any indicators of displacement. They use parcel sales data that can only show the rising housing values, which is not a direct indicator of displacement. In fact, there may be little direct link between an increase in housing *value* and an increase in housing *cost* if effective local property tax policies such as property tax caps are in place.

For instance, New York City limits the assessed value of Class 1 property (i.e., one-to-three-unit, predominantly residential properties) at a maximum of 6% per year or 20% over five years and the assessed value cannot exceed 6% of the home's market value (NYC Department of Finance, 2023). A home sold for a million-dollar market value would have an assessed value of 6% of the market value, or $60,000, in the first year. In the second year, even if the house's market value is appreciated by 20% for any reason, the assessed value can increase only up to 6% per year, or to $63,600 at the maximum. Black and Richards (2020) estimate that homes closest to High Line Park experienced a housing price increase of 35%, but this should not mean a 35% increase in housing cost for existing homeowners. Because of the property tax cap, the increase in housing cost for homeowners is a maximum of 6%. The 6% increase in housing costs is not trivial, but it means there could be a substantial difference between the value of a property and the actual costs the homeowner is responsible for. Without understanding this policy environment, the assumed link between an increase in housing *value* and an increase in housing *cost* likely overstates the risk of displacement of existing homeowners because the homeowners' property tax rates will not mirror the actual value of the property—the rates could be substantially below market value.

In addition, New York City is known for its rent regulation, such as rent control and rent stabilization laws. Approximately 16,400 rental units are subject to rent control, and 44% of all rental properties in New York City (NYC) are under rent stabilization (New York City Rent Guidelines Board, n.d., para 4).[1] The NYC Rent Guideline Board meets every year and sets annual rent increases. For instance, it set a 3.25% increase for 1-year lease agreements and a 5% increase for two-year lease agreements, which applied to all leases signed between October 2022 and September 2023 (New York City Rent Guidelines Board, 2022). Thus, when units are under rent regulations, renters see the regulated rent increase each year, even in areas where the housing prices soar due to environmental improvement. Once again, the actual rent that a renter pays could be substantially below the rent for the same unit if it were on the rental market.

The property tax caps and rent regulations in New York City apply to all residents regardless of their income levels. In addition, the State of New York implements an income-based housing cost relief program known as a circuit-breaker (Bowman et al., 2009) that reduces property taxes or rents for people with lower or moderate incomes. When someone's property tax or rent is more than a certain percentage of their income, the circuit-breaker lowers their taxes or rental rates to make their home more affordable. This way, people are less likely to pay more taxes or rent than they can afford, especially in places where housing values rise quickly.

[1] Rent regulation includes rent control which freezes rent payment and rent stabilization which allows landlords to raise rents by a certain percentage each year. There are about 2 million renter-occupied housing units in New York city (NYC, 2024).

While some states make only homeowners or renters of certain ages eligible for such programs, New York allows both homeowners and renters of all ages. Although the state's circuit-breaker program has been under consideration for an overhaul for its inadequacy (Bowman et al., 2009), it is another layer of protection that homeowners and renters can benefit from, in addition to property tax caps and rent regulation.

The New York City example suggests that we cannot rely solely on changes in property values to understand the extent of displacement that takes place, because real property values may not reflect the actual costs to the residents that live on those properties. Researchers agree that the nature and magnitude of displacement need to be better quantified (Zuk et al., 2015). The first step is to understand the local policy environment for homeowners and renters because, as explained in the case of New York City above, changes in housing costs due to green gentrification differ for homeowners and renters, depending on state and local policies.

One of the noteworthy aspects of these policies is that both state and local governments may set the laws for property taxes and rent regulation, which leads to highly heterogeneous policy environments and therefore, potentially, heterogeneous impacts on green gentrification. Table 8.1 outlines state measures aimed at preventing homeowners and renters from being forced out due to increases in property taxes or rents, including property tax assessment limits and rent regulation policies for all income levels as well as circuit-breaker programs for low-income homeowners and renters.

The state policy environment detailed in Table 8.1 offers an array of policy choices. However, the list of options provided is not exhaustive. There are many other programs for property tax relief available at the state and local levels. The specifics can differ significantly, with even more variations at the local level, particularly in cities. The diverse policy environment means that no single policy can universally ensure protection from displacement. To illustrate, in Table 8.2 we show information for large US cities in several states, presenting their housing-price-to-income ratios (higher values indicating lower housing affordability), property tax laws, and rent regulations that directly affect displacement.

Homeowner Protections

Assessment limits, or property tax caps, are well-known property tax policies that are intended to protect homeowners from displacement due to high property tax bills. Many major cities, including New York City, Los Angeles, Chicago, Houston, and Phoenix, implement some version of a property tax cap. These policies are explicitly designed to stabilize neighborhoods, and although many previous studies have examined cases in these cities, they have rarely been considered in the literature related to green gentrification.

As of 2023, 16 states and DC implement property tax assessment limits that protect homeowners from housing market volatility. The tax literature generally agrees that property tax caps incentivize homeowners to extend their housing tenure

Table 8.1 State-level policies designed to protect homeowners and renters from displacement

State	For homeowners		For renters	
	Assessment limits (maximum %)	Circuit-breakers for low-income homeowners	Renter protection (rent control, rent stabilization laws)	Circuit-breakers for low-income renters
Alabama	–	–	Preempts local	–
Alaska	–	–	No legal language	–
Arizona	10%, all residential	Yes (65+, disabled)	Preempts local	Yes (65+, disabled)
Arkansas	5%, owner-occupied	–	–	–
California	2%, all residential	Yes (62+ or disabled)	State and local	Yes (62+ or disabled)
Colorado	–	Yes (65+, disabled)	Preempts local	Yes (65+, disabled)
Connecticut	Local option	Yes (65+, disabled)	Preempts local	Yes (65+, disabled)
Delaware	–	–	No legal language	–
District of Columbia	10%, owner-occupied	Yes (all ages)	District-wide	Yes (all ages)
Florida	3%, owner-occupied	–	Preempts local	–
Georgia	Freeze for age 65+ owner-occupied	–	Preempts local	–
Hawaii	–	–	No legal language	–
Idaho	–	Yes (65+, disabled)	Preempts local	–
Illinois	Local option, owner-occupied	Yes (65+, disabled)	Preempts local	Yes (65+, disabled)
Indiana	–	–	Preempts local	–
Iowa	–	Yes (65+, disabled)	Preempts local	Yes (65+, disabled)
Kansas	–	Yes (55+, disabled)	Preempts local	Yes (55+, disabled)
Kentucky	–	–	Preempts local	–
Louisiana	–	–	Preempts local	–
Maine	–	Yes (all ages)	Local only	Yes (all ages)
Maryland	10%, owner-occupied	Yes (all ages)	Local only	Yes (all ages)
Massachusetts	–	Yes (65+)	Preempts local	Yes (65+)
Michigan	5%, all residential	Yes (all ages)	Preempts local	Yes (all ages)
Minnesota	–	Yes (all ages)	Local only	Yes (all ages)
Mississippi	–	–	Preempts local	–
Missouri	-	Yes (65+ or disabled)	Preempts local	Yes (65+ or disabled)
Montana	16.66% over 6 years, all residential	Yes (all ages)	Preempts local	Yes (62+ or disabled)

(continued)

Table 8.1 (continued)

State	For homeowners		For renters	
	Assessment limits (maximum %)	Circuit-breakers for low-income homeowners	Renter protection (rent control, rent stabilization laws)	Circuit-breakers for low-income renters
Nebraska	–	Yes (65+, disabled)	No legal language	–
Nevada	–	Yes (62+)	No legal language	Yes (62+)
New Hampshire	–	Yes (all ages)	Preempts local	–
New Jersey	–	Yes (all ages)	Local only	Yes (all ages)
New Mexico	3%, all residential	Yes (65+)	Preempts local	Yes (65+)
New York	Local only (e.g., New York City and Nassau County, 6–8%, residential properties < 8 units)	Yes (all ages)	Local only	Yes (all ages)
North Carolina	15% over 5 years for all residential	–	Preempts local	–
North Dakota	–	Yes (65+, disabled)	Preempts local	Yes (65+, disabled)
Ohio	–	–	Preempts local	–
Oklahoma	5%, all residential	Yes (65+, disabled)	Preempts local	–
Oregon	3%, all residential	–	State and no local	Yes (58+)
Pennsylvania	–	Yes (65+ or disabled)	No legal language	Yes (65+ or disabled)
Rhode Island	–	Yes (all ages)	No legal language	Yes (all ages)
South Carolina	15% over 5 years, owner-occupied	–	Preempts local	–
South Dakota	–	Yes (65+, disabled)	Preempts local	–
Tennessee	–	–	Preempts local	–
Texas	10%, owner-occupied	–	Preempts local	–
Utah	–	Yes (65+)	Preempts local	Yes (65+)
Vermont	–	Yes (all ages)	No legal language	Yes (all ages)
Virginia	–	–	No legal language	–
Washington	–	Yes (61+ or disabled)	Preempts local	Yes (61+ or disabled)
West Virginia	–	Yes (all ages)	No legal language	–
Wisconsin	–	Yes (all ages)	Preempts local	Yes (all ages)
Wyoming	–	Yes (65+, disabled)	No legal language	Yes (65+, disabled)

Sources: Assessment limits from Lee (2023), Circuit-breakers from Bowman et al. (2009), Rent regulation from National Apartment Association (2023)

(continued)

Table 8.1 (continued)

Notes: *Preempts local*: the states do not have a state-wide regulation and preempts/prevents local rent regulation
No legal language: the states have no legal language about rent control
State and no local: Oregon has state-wide rent control but preempts local governments from adopting additional rent regulations
State and local: California has state-wide rent control, and local governments can adopt additional rent control measures

(Wasi & White, 2005).[2] They restrict the annual increase in property tax, so homeowners benefit from more stable and predictable property tax bills over time. For instance, California's assessment limit, known as Proposition 13, caps the annual property tax increase to a maximum of 2% from the previous year, regardless of the housing value appreciation of that year. Public finance literature has established that the assessment limits in California provide substantial property tax subsidies to homeowners regardless of households' socioeconomic and demographic characteristics (Wasi & White, 2005).

Even without state-wide assessment limits, some local governments implement homeowner protection measures. For instance, Philadelphia adopted two property tax relief programs, the low-income senior citizen Real Estate Tax freeze program and the Longtime Owner Occupants Program (LOOP). These programs allow eligible homeowners to freeze property taxes to prevent future tax increases. According to the National Community Reinvestment Coalition, eligible homeowners saved $21 million in property taxes from these programs (McConnell, 2019). Charlotte, Seattle, and Denver do not have property tax assessment limits, but they have circuit-breakers to protect low-income senior or disabled homeowners (Bowman et al., 2009). Eighteen states and DC have circuit-breaker programs for low-income individuals. The programs provide financial assistance when property tax bills exceed a certain level of income threshold. They significantly alleviate the property tax liability for eligible homeowners, making it more affordable for them to maintain housing stability. In Table 8.2, Columbus and Indianapolis are the only two cities without any property tax relief program for homeowners. However, their price-to-income ratio is under 4.0, meaning residents spend less of their income on housing, which indicates lower displacement risks relative to other housing markets. Income thresholds vary across states in that some jurisdictions (e.g., New York and DC) make the relief available only for very low-income households whose income is less than $30,000 per year. Other states, such as Michigan and New Jersey, allow households with income above $70,000 also to be eligible for the program.

[2] "Housing tenure" is how long people remain in the same residence.

Table 8.2 Selected populous cities in the U.S. and policies protecting residents

				For Homeowners		For Renters	
City	State	2022 population	House Price-to-income ratio (housing affordability)	Assessment limits (property tax cap, maximum %)	Circuit-breakers for low-income homeowners	Renter protection (rent control, rent stabilization laws)	Circuit-breakers for low-income renters
New York	NY	8,335,897	10.5	Yes (6%)	All ages	Yes (local)	All ages
Los Angeles	CA	3,822,238	13.1	Yes (2%)	Age 62+ or disabled	Yes (local and state)	Age 62+ or disabled
Chicago	IL	2,665,039	4.9	Yes (5%)	Age 65+ or disabled	No (preempt)	Age 65+ or disabled
Houston	TX	2,302,878	4.5	Yes (10%)	None	No (preempt)	None
Phoenix	AZ	1,644,409	6.2	Yes (10%)	Age 65+ or disabled	No (preempt)	Age 65+ or disabled
Philadelphia	PA	1,567,258	4.8	No	Age 65+ or disabled	No (no preempt)	Age 65+ or disabled
Jacksonville	FL	971,319	4.5	Yes (3%)	None	No (preempt)	None
Columbus	OH	907,971	3.8	No	None	No (preempt)	None
Charlotte	NC	897,720	5.2	No	Age 65+ or disabled	No (preempt)	None
Indianapolis	IN	880,621	3.9	No	None	No (preempt)	None
Seattle	WA	749,256	8.5	No	Age 61+ retired or disabled	Yes (local)	None
Denver	CO	713,252	7.5	No	Age 65+ or disabled	No (preempt)	Age 65+ or disabled
Oklahoma City	OK	694,800	3.1	Yes (5%)	Age 65+ or disabled	No (preempt)	None
Washington D.C.	.	671,803	3.9	Yes (10%)	All ages	Yes (district-wide)	All ages

Notes: Rent regulation includes rent control or rent stabilization laws. Indirect regulations such as eviction or repair/maintenance regulations are not included; "preempt" indicates that state law preempts local law

Sources: Population (U.S. Census), Price-to-income ratio (median home price to median household income, U.S. Census), Assessment limits (Lee, 2023), Circuit-breakers for homeowners and renters (Institute of Taxation and Economic Policy. 2023. "Report: Preventing an Overload: How Property Circuit Breakers Promote Housing Affordability."), Renter protection (National Apartment Association, 2023. "Rent Control Statutory Insights")

When considering these property tax policies and relief programs, one expects that the conventional narrative likely overestimates the increase in housing costs and the risk of displacement of homeowners due to green projects. In fact, in many cases, homeowners may benefit from better green amenities and increased home equity without the risk of losing their property due to the property tax increase. On the other hand, it is possible that, without these local policies, the observed magnitude of displacement may underestimate the forces of green displacement, as the policies could have protected residents who may have otherwise been displaced. Therefore, the magnitude of displacement without those policies could have been higher.

Renter Protection

Existing studies often recommend housing rights laws such as rent regulation, housing trusts, or social housing to prevent green displacement (Anguelovski et al. 2022). However, as Table 8.2 shows, homeowner-protective measures are much more prevalent than policies to protect renters. Only seven states and DC have local- and/or state-level (California and Oregon) rent regulation, while more than 30 states prohibit local governments from enacting such measures, according to the National Apartment Association (2023). This uncommonness of the policy may support the observation that green projects tend to displace renters (Zuk et al., 2015). Table 8.2 shows that cities that have tended to adopt rent regulation have strong housing markets with a price-to-income ratio over 8 (Los Angeles 13.1, New York 10.5, Seattle 8.5).

It is important to note that rent regulation does not necessarily protect all low-income renters because what is controlled is the rental unit, which may or may not be occupied by low-income households. In other words, rent regulations are generally not means-tested and thus not targeted to protect low-income renters, although obviously many will be protected anyway. To protect specifically low-income renters, some states or local governments extend circuit-breakers to renters (Bowman et al., 2009). Those states recognize that even if renters do not directly pay property taxes, they can still be affected by property tax increases that are passed on to them through higher rents. Thus, their circuit-breaker laws also protect renters from property-tax-related rent increases. It is hard to know exactly what portion of the property tax landlords charge renters, but the states estimate that it's between 6% and 25% (Lyons et al., 2007). Table 8.2 shows that New York, Los Angeles, Chicago, Phoenix, Philadelphia, and Denver implement circuit-breakers for renters to relieve the overload of housing costs in their budgets. In most places, low-income seniors and disabled individuals are eligible, but New York City allows all ages to be eligible to apply (Bowman et al., 2009).

Importance of Policy Environment in Assessing Displacement Driven by Green Projects

These local policies are critical in assessing the displacement risks of residents. For example, one of the types of green projects that impacts the nearby housing market is urban greenway projects (Rigolon et al., 2020). The successful High Line Park project in New York City has become a model for Boston's Rose Kennedy Greenway, Philadelphia's Rail Park, DC's 11th Street Bridge Park, Chicago's The 606, Atlanta's BeltLine, Los Angeles' River Revitalization, and others. However, the impact of these projects on displacement can only be understood within the context of their local policies for homeowners and renters.

In Los Angeles, the risks for homeowners' displacement due to the increase in housing values are expected to be lower because the assessment limit caps the housing cost increase at the maximum of 2% for all homeowners, and the circuit-breaker program for homeowners adds an extra layer of protection for low-income homeowners. Los Angeles also implements rent-stabilization and circuit-breaker programs for low-income senior and disabled renters. In Denver, low-income senior or disabled homeowners and renters can access the circuit-breaker programs. However, there is neither an assessment limit nor rent regulation, which allows the increased housing value to raise the housing cost. Thus, the risk distributions for homeowners and renters differ between the cities and within the cities depending on the policy combination and local homeownership rates.

Some researchers advise anti-displacement measures for neighborhood stabilization. The Center for American Progress recommends renter protection measures without mentioning homeowner protection (Dorazio, 2022). Nevertheless, rent control, rent stabilization, anti-eviction protections, or renter education workshops cannot be adopted locally in states that preempt local adoption of rent regulation and this restriction is rarely recognized (Immergluck & Balan, 2018). The Lincoln Institute of Land Policy encourages local officials to explore income-based property-tax relief, subsidized rental costs, capped property taxes for long-time residents, and housing land trusts (Immergluck & Balan, 2018; Schilling et al., 2023). Local governments do implement some parks-related anti-displacement strategies or PRADS, according to the National Recreation and Park Association (Rigolon & Christensen, 2019). In the study of 28 park development projects in 19 cities, 13 park projects in 12 cities employed 26 types of PRADS.

To summarize, green-project-driven residential property value appreciation, property tax relief programs, and tenant protection laws vary significantly across states and localities and are further influenced by population demographics, housing market conditions, project characteristics, and the initial neighborhood profile. These factors shape the extent and nature of displacement risks associated with green gentrification. The distribution of displacement risks differs for homeowners and renters, depending on local property tax laws and tenant protection regulations. The risk distribution between homeowners and renters will determine the overall risk of displacement in the neighborhood after environmental improvements lead to property-value increases—and attributes of those who are displaced. Therefore, successful policies in one locality only sometimes succeed in another locality. It is imperative to be aware that anti-displacement policies intended to ensure environmental justice interact with existing policies to determine the nature and degrees of displacement risks.

Heterogeneous policy environments are a crucial dimension to understanding the processes of gentrification and displacement driven by environmental improvements. With understanding these factors, attempts to address the challenges of green gentrification and displacement risks in urban areas may provide practical and more realistic policy tools. While acknowledging the genuine concerns and challenges of green gentrification and displacement, in this chapter we explore a more nuanced and context-driven approach to research and policymaking.

Agent-Based Modeling

Here we use variations on the ABM described in previous chapters to assess the impact of these policy choices within different contexts. The model required several changes in order to assess the impact of property tax caps and rent regulation measures. We use the same base model described in detail in Chap. 5 and also include the environmental improvement policies described in Chap. 7. We do not elaborate on the model here, but a full description of all the model's behavior can be found particularly in Chap. 5. By adding to the same model we used before, we can assess outcomes when we implement these additional policy measures designed to ameliorate displacement.

Base Model Similarities and Differences (Chaps. 5, 7, and 8)

We still base the ABM model used in this chapter on the real estate market described in Chap. 5, and the environmental policy improvements that spur gentrification as described in Chap. 7. However, given the policies implemented in this chapter, the model needed to be adjusted in two key ways. First, the model in this chapter differentiates between renters and homeowners, while in our previous ABM models all residents are homeowners (though we discuss implications for renters). Secondly, this model considers property taxes or rising property tax assessments that might displace homeowners, while we only considered house-price changes in the previous model, abstracting from what factors cause price changes. Previously, we had not differentiated between the actual *value* of a plot and the *costs* of staying on that plot for residents.

Other than the two revisions described below, the model functions the same as the version in Chap. 7 with one simplification. Rather than explore the park placement policies broadly as we did in Chap. 7, in this chapter we only consider two of the scenarios from Chap. 7: the just green enough (JGE) scenario (referred to as S-LP-LD in Chap. 7) and its counterpart where environmental improvement is the focus and gentrification is not much of a consideration, which here we refer to as the environmental focus, or EF scenario (referred to as L-NP-HD in Chap. 7). The first scenario places many small parks, of lower desirability, in the lowest wealth areas,[3] while the second scenario places a few large parks, with high desirability, in the most polluted areas. This allows us to compare the effect of property tax and rental cap policies in scenarios where avoiding gentrification is the key focus versus scenarios where environmental improvement is the focus.

We then do a simple comparison of the presence or absence of property-tax caps and/or rent-increase caps. We set both at 2%, meaning that policy does not allow property tax rates, nor rents, to increase by more than 2% per tick. We opted for a relatively low cap in order to ensure that we capture the effect of the policies in a widespread manner. For reference, prices of plots increase by about 3-5% per tick, depending on the scenario, so a 2% cap ensures that most residents will be affected by the policy. We limit the caps to those patches in the model that hit the gentrifiable threshold at some point, i.e., the policies are only in effect in those areas that at some point were amongst the poorest of the region. We therefore treat the cap policies as circuit-breakers, only available to those who meet certain criteria—in this case, the sole criterion is being located in an area that is gentrifiable, which ensures that our policies are only applicable to poor areas. We recognize this as a relatively narrow implementation of these policies, but, for purposes of our model, we believe that a more targeted policy approach makes sense owing to the relatively confined environment of the model. A policy that is applicable across the board would fundamentally alter our housing market, making it hard to distinguish the effect of policy interventions, whereas the narrower approach we have taken ensures that we can isolate the effect of the policy more clearly.

Homeowners Versus Renters

To differentiate between homeowners and renters, we created a new variable indicating whether a resident is a renter or homeowner. We align the probability that a resident is a renter in terms of wealth, roughly approximating trends observed in US census data, with about 1/3 of the resident population renting at any given time, and 2/3 owning, and renters being poorer on average than homeowners. Residents draw a random number between 0 and 1 and round to either 0 or 1, with 1 indicating a renter and 0 an owner. The mean of the distribution from which each resident draws is different depending upon their wealth level (standard deviations are the same for each, at 0.25). Residents with a wealth value in the lowest quartile draw their

[3] Because of interconnections in the model, low-priced land is almost inevitably in low-wealth neighborhoods.

random number from a distribution with a mean of 0.6, indicating that we would expect a high proportion of renters in this lowest quartile. Residents in the top wealth quartile draw from a distribution centered at 0.1, indicating that we expect relatively few renters in this category. Residents in the middle two quartiles draw from a distribution centered at 0.35.

We do not differentiate by race in the likelihood of being a renter or homeowner. However, note that the minority residents are poorer on average than the majority residents. Thus, we always end up with more minority renters (42-45% of minorities are renters), than majority renters (30% of majorities are renters), and more majority homeowners than minority homeowners.

Since we conceive of property affordability based on wealth, we do not differentiate the home search process between renters and owners. Residents do not pay rents or mortgages in our model. They simply seek places to live on the basis of their utility functions and constraints, including the budget constraint. The key difference between renters and owners in the model is that when owners relocate during the regular operation of the model, their wealth is adjusted based upon the price differential between where they moved from to where they moved to (thus wealth can increase or decrease), whereas renters' wealth does not change when they move. This is important as we consider the implications of displacement; we consider displacement the same as in previous chapters: residents are displaced, i.e., forced to move to a different plot, when the value of the plot they are currently located on exceeds their budget constraint. This means that when owners are displaced from gentrifying or gentrified areas, their wealth increases by the amount they "sell" their current plot for, less the price of the plot they move to. Renters' wealth is not adjusted when they are displaced. In gentrifying or gentrified areas, then, owners realize wealth gains when they are displaced, and renters do not.

Tax Caps

The second key change has to do with taxes, or the additional costs of properties. Our previous versions of the model did not include taxes, largely because we assumed taxes to be included within the overall price of a plot. Since we are now differentiating between renters and owners, we need to differentiate between the mechanisms that might lead to displacement for each group. Rather than create taxes, per se, we opted to handle plot price increases differently for each group. Essentially, we need to operationalize a means through which residents can avoid forced displacement, even if the price of the plot they are located on exceeds the upper bound of their budget constraint. This is effectively what both rent control and property tax abatement policies do – they enable people to remain where they are by artificially reducing the cost of a plot for the residents who live there, as the increased value of the land they are occupying does not necessarily increase the cost that they actually pay to remain on that land.

For renters, this process is straightforward, both in the real world and in the model. In the real world rent controls protect renters from rapid upswings in market prices by limiting the amount that a renter's rent payment can increase per year. If a policy provides for a 2% growth cap on rent, this means that a current leaseholders' rent cannot increase by more than 2% regardless of what the market might bear for a new lessee. Such a cap could result in residents in the same facility paying different rents, with a longtime renter perhaps paying substantially less in rent than someone renting in the facility for the first time. We handle this similarly in the model.

In previous versions of the model, we did not differentiate between the price of a plot and the resident's rent/mortgage because we had no policy that controls cost escalation. In this chapter, we track plot prices as we did before, but now also track resident costs to live on the plot. We implement rent caps much as in the real world process described above. If a renter is located on a plot that is realizing value increases, a renter's rent can only increase by the maximum that the cap allows per tick. For example, if the rent cap is set at 2%, the renter's rent cannot increase by more than that. If the plot's price is decreasing, or increases by less than 2%, then the renter's rent is adjusted according to the actual market price. However, if the plot price increases by more than 2% in a given tick, the rent for the resident will only increase by 2%. If, in subsequent ticks, the plot's price increases even further, the renter's actual price can still only increase by a maximum of 2%. This means that the market value of a plot and the price that is paid by an existing resident can diverge relatively quickly. A resident is no longer displaced when the plot price exceeds their budget constraint but is now only displaced when their actual housing *cost*, limited by the rental cap, exceeds the budget constraint. That is, we would expect the cap to prolong tenure and the time until displacement occurs. Residents can obviously still move voluntarily, or they can be displaced, and when they are, any new residents that are located on the plot pay the full market price of the plot. Similarly, when a renter leaves a plot, either voluntarily or because of displacement, they must pay the market rate of their new location (although they could still be eligible for rental caps if they move into a gentrifiable area).

For homeowners, we handle things similarly, but with some key differences. Since we do not incorporate property taxes in the model, we again conceive of the difference in the actual cost of a plot and an owner-resident's realized cost. In the real world, property tax abatements function like rent caps, in that they limit the amount that a property owner's taxes can increase in a given year. In our model, we factor this into the value of the plot, the same as we do with rent caps. If we set the property tax cap at 2%, this means that owners, like renters, cannot see the realized cost of a plot rise more than 2% per tick, regardless of the actual increase in the cost of the plot they are located on. The key difference between homeowners and renters is that, as homeowners experience costs that are lower than the actual value of the plot, they are accruing equity. When a renter is displaced or chooses to move from a plot whose value is increasing, they may simply move to a more affordable plot with no gains realized. When an owner moves, they receive the actual value of the plot they move from. So, if an owner has a 2% property tax cap, their personal costs of a plot cannot increase by more than 2% per tick, meaning they cannot be displaced until their realized costs (rather than the actual price of the plot) exceeds their existing budget constraint, assuming they do not move. However, whereas this cap incentivizes renters to stay put, it is more complicated for owners as they also take into account the potential wealth increase, they could accrue by moving, because if they "sell" and relocate, their return is based on the market price of the plot, not on the artificially lowered costs they are spending to live there.

So, functionally the caps work similarly for both renters and owners, but the effect of the cap is different, and there is an important functional difference applied to the budget constraint. Neither owners nor renters are displaced unless their realized cost exceeds their budget constraint, and the budget constraint for renters does not change because we assume that they do not accrue wealth through a home sale. The budget constraint for homeowners can change, however. They are displaced if and when the realized cost of the plot they are on exceeds their budget constraint and are forced to move, but they still realize wealth gains based upon the actual price of the plot they are vacating (less the price of the plot they move to). In effect, then, owners' upper-bound budget constraint should include the potential difference between the plot they are on and any plot they may move to. Therefore, when owners consider moving, their upper-bound budget constraint is assumed to be the market price of the plot they are on, because if they do relocate, they can realize these gains. What this functionally does is allow owners to consider relocation to different plots that exceed their existing budget constraint prior to gentrification, under the assumption that, if they move, their wealth will increase by that price differential.

As noted above, the model otherwise operates the same as in previous chapters. Our outcomes of interest are the extent to which gentrification-related displacement occurs and how long the process takes to play out. Our scenarios, in this case, relate to different policy choices with respect to the presence or absence of rent caps and property-tax caps and how different scenarios of land-use choices interact with the use of rent and property tax control policies. Last, after performing our base analyses as described her, we also create some configurations that approximate specific cities of interest and see how city-specific variable clusters affect our findings.

Simulation Scenarios

Table 8.3 shows the general scenarios that we ran. As noted, we simulated two scenarios from Chap. 7, the JGE and EF settings, and varied the utilization of 2% caps on both rent increases and property tax increases. The various combinations of these policies resulted in 8 different scenarios.In addition, drawing from the information in Table 8.2, we ran an additional set of trials varying the contexts by approximating circumstances similar to six different cities, as depicted in Table 8.4. Since our model is not reflective of an actual population, we compare the cities in a relative manner. Population size starts at 2000, with Denver and Seattle serving as the base cases (because they have similar city populations) and the other cities starting with

Table 8.3 Policy scenarios simulated

Just green enough (JGE) scenarios			Environmental focus (EF) scenarios		
	S-LP-LD			L-NP-HD	
Scenario 1–4	Rent cap	Tax cap	Scenario 5–8	Rent cap	Tax cap
Scenario 1	⊗	⊗	Scenario 5	⊗	⊗
Scenario 2	**P**	⊗	Scenario 6	**P**	⊗
Scenario 3	⊗	**P**	Scenario 7	⊗	**P**
Scenario 4	**P**	**P**	Scenario 8	**P**	**P**

Key:
First letter: S—Small parks scenario; L—Large parks scenario
Second set of letters: LP—Lowest Price plot chosen; NP—Near Pollution (most polluted) plot chosen
Third set of letters: LD—Low Desirability parks placed; HD—High Desirability parks placed
⊗: condition absent (no cap)
P: 2% cap Present

Table 8.4 City approximations and variable settings

City	Population size (at start)	Growth rate	Density	Housing burden tolerance	Rent cap	Property tax cap
Denver	Base	6%	4	0.5	None	None
Seattle	Base	5%	3	0.6	5%	None
Indianapolis	15% higher	2%	2	0.3	None	None
Chicago	300% higher	3%	3	0.4	None	5%
New York	850% higher	4%	6	0.7	5%	10%
Los Angeles	400% higher	6%	7	0.8	5%	2%

populations proportionally larger than this base case. Growth rates are approximations of the average growth over the past 20 years for each city. We do not make adjustments for the geographical characteristics of these cities, and although we did adjust the demographic characteristics to reflect the real cities, our two-race model is obviously just a rough approximation of these cities. The city comparisons should thus be taken as illustrative rather than literal.

To approximate some important city-based conditions, we utilize an existing variable, density, and add a new one, housing burden. Based on the empirical information in Table 8.2 and also our previous research, particularly the results in Chap. 5, our expectation is that there may be differences in how these policies play out depending upon how dense a city is, and the extent to which people are willing to tolerate more or less of their wealth being spent on housing (as shown in Table 8.2 this varies across US cities, with residents of some cities highly housing burdened as defined by the US government). For the density variable, we simply adjusted the mean density preference from which residents randomly drew their individual density preferences, to reflect the densities of the cities shown in Table 8.2. As noted above, we coupled this with either smaller or larger starting populations to reflect the city's size, and either smaller or larger growth rates to approximate each cities' growth pattern over recent decades.The new variable, housing burden, functions as a second means of considering the budget constraint. For this set of trials, we altered the budget constraint that we used above. As noted in previous chapters, our wealth variable is drawn from yearly income data from the US census, and our housing values are drawn from median home values, also from the US census. We considered the budget constraint for plot affordability from this perspective – i.e., the upper bound of the budget constraint for a plot is set at three times a resident's wealth. To determine the housing burden, we created a ratio of each resident's wealth and 1/3 of the home value, to roughly approximate a one-year housing burden measure. The housing burden variable is, therefore, a value between 0 and 1 that represents the proportion of a resident's wealth that they spend per tick on housing.We use this housing burden variable as an approximation of housing cost differences and the willingness of people in different settings to spend more of their income on housing. In these sets of trials, each individual draws a unique housing burden tolerance from a normal distribution with a standard deviation of 0.05, centered at a different mean, depending on the city that is being modeled. This leaves us with six different cities, with six different densities and housing burden tolerances, and a unique mix of cap policies.

Results

We illustrate our results in the Figures below, showing our key variables between ticks 27 and 60 as we do in Chap. 7. Just as in Chap. 7, we use an initial 25-tick setup to ensure that we have a world with a high level of segregation and environmental injustice. After tick 25, we settle in with a relatively conservative estimate of segregation preferences of 25% for both groups of residents, as we do in Chap. 7. Broadly speaking, the results point to two main themes: First, virtually any policy effort, whether using a JGE approach or implementing any combination of property tax caps or rent caps, results in less harmful displacement effects than simply following an environmental focus (EF), and secondly, the benefits of the policies are received almost entirely by the minority residents rather than the majority.

As demonstrated in Fig. 8.1, majority residents are displaced at roughly the same rate regardless of the policy approach. Minority residents are displaced at a higher overall rate than majorities, but there is a clear distinction in the displacement rate when following the EF with no caps approach (the brighter green line) compared to following any other approach. Further, the rates follow different patterns. In the outset of policy implementation, about 10% of majority residents are displaced, and the rate steadily rises to about 30%. The initial displacement of minority residents is lower, likely owing to our initial setup conditions of creating a highly segregated society, but the climb throughout the trials is higher for minority residents, starting at zero and rising to nearly 40% in the EF with no caps scenario.

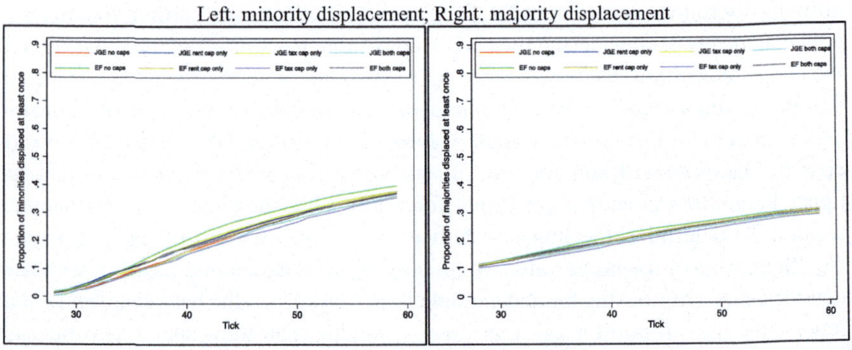

Fig. 8.1 Rates of displacement at least once for both resident types

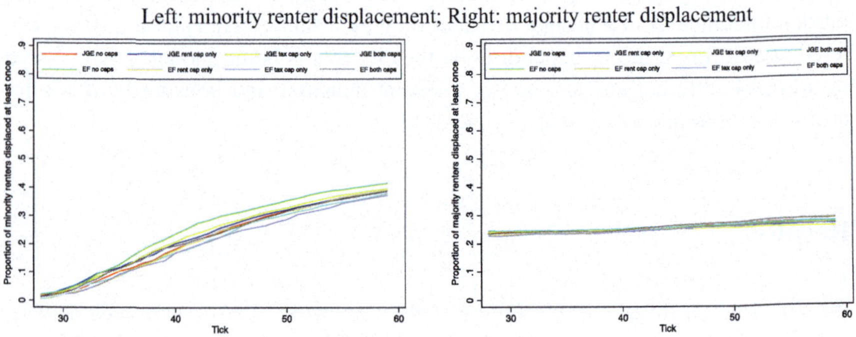

Fig. 8.2 Rates of displacement at least once for renters of both types

If we narrow the focus to look more specifically at majority renters versus minority renters, as in Fig. 8.2, we see a similar result. Again we see that majority renters are quickly displaced as we relax the initial highly segregated assumption, but the overall rate of majority renter displacement remains stable throughout time (around 25%). In contrast, displacement of minority renters in Fig. 8.2 follows a similar trend to the displacement of minority residents overall in Fig. 8.1, with the EF with no caps scenario clearly leading to more displacement of minority renters than any of the other scenarios.

Turning to homeowners, we see trends very similar to those for renters, with a couple of caveats. Once again, the displacement of majority homeowners is stable over time and not influenced by policy choices, but majority homeowners are only displaced at about a 10% rate, whereas majority renters see steadily rising rates with displacement ending at about a 35% rate. Similarly, the overall trend of minority homeowner displacement is similar to minority renter displacement. Once again the EF with no caps approach leads to the most overall displacement, while the other scenarios are clustered closely together, and, similar to the situation for majority renters, the overall level of majority homeowner displacement is lower than minority homeowner displacement, although the trend lines are closer for minority residents, suggesting that home-ownership is less protective for minorities than for majorities. The peak level of minority renter displacement is about 40% at tick 60, whereas the peak level of minority homeowner displacement is about 35% (Fig. 8.3).

In Fig. 8.4, we consider the tenure rates, or the amount of time that residents on average remain at a plot, for renter residents of both types. We include only renter tenure here as owner tenure was similar in both cases, with homeowner tenure averaging around 3 for both types of residents regardless of the scenario. Renter tenure was slightly more varied, at least for minority residents, although trends were difficult to discern. Majority renter tenure was not affected by the scenario at all, clustering closely at about 1.1 ticks and remaining at that level throughout the trials. Minority tenure started higher overall than majority tenure, and, although declining throughout the trials for each scenario, stayed higher. The result for minority renter tenure is consistent with what we see in the previous graphs, with lowest tenure being found in the EF with no caps scenario, and the highest tenure (from about tick 35 to 55) associated with the two scenarios (one JGE and one EF) in which there are

Left: minority homeowner displacement; Right: majority homeowner displacement

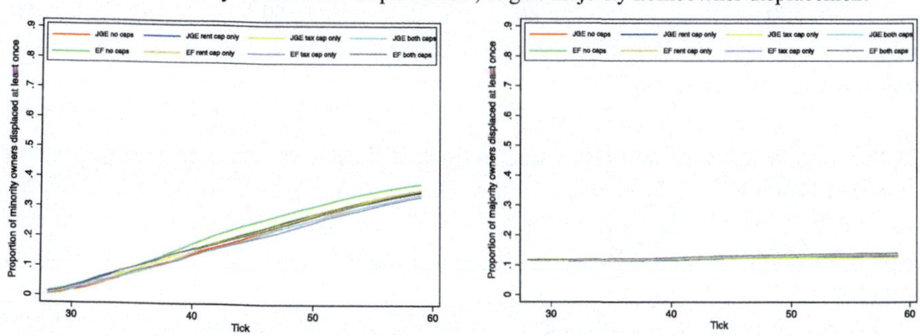

Fig. 8.3 Rates of displacement at least once for homeowners of both types

Left: minority renter tenure; Right: majority renter tenure

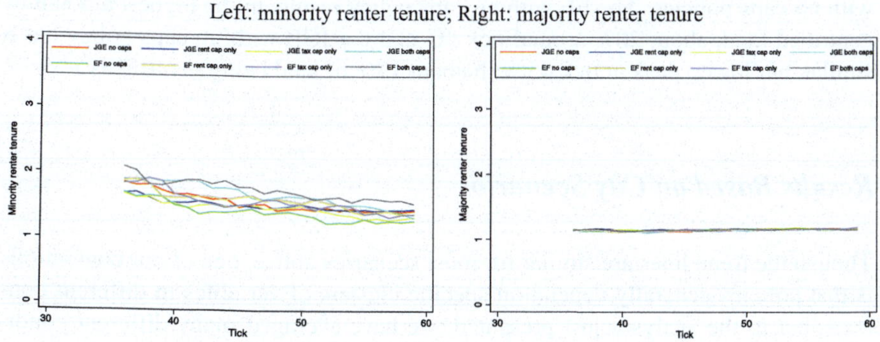

Fig. 8.4 Tenure for renters of both types

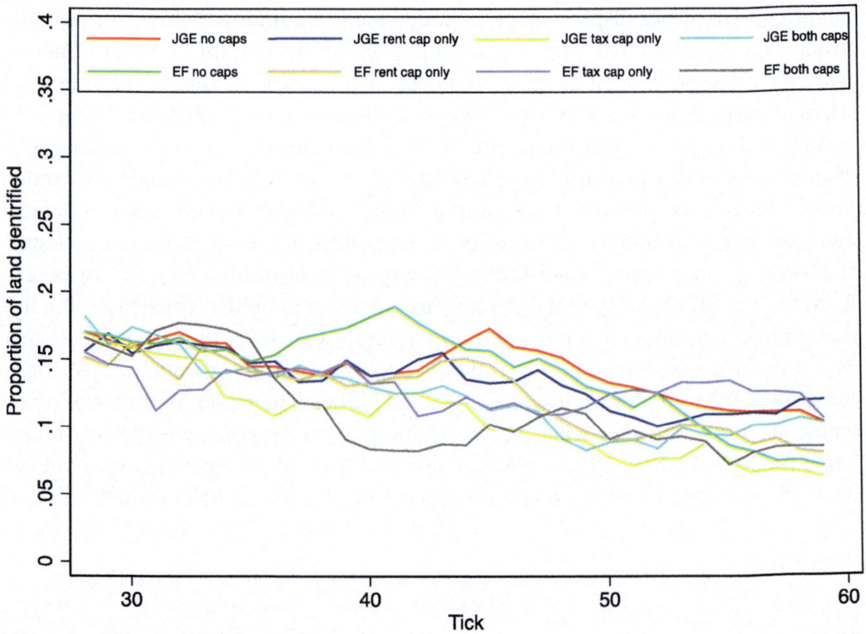

Fig. 8.5 Proportion of gentrified land

both property tax caps and rent caps. In both of these cases, minority renter tenure hovered around 1.75, in contrast to the EF with no caps scenario which hovered around 1.25. While this difference may not seem like much, translated into real time, this could mean a difference of about 6 months—averaged over all minority renters.

Finally, in Fig. 8.5, we consider the results from a land-use perspective, in terms of the amount of land that gets gentrified during each of the scenarios. We see a relatively stable level of land that is gentrified, between about 5% and 17%. However, there are two clear peaks of land that are gentrified, both occurring under scenarios with no-caps policies. The EF with no caps policy results in the highest amount of gentrified land, about 20% around tick 40, and the JGE with no caps policy has a similar, but lower, peak at tick 45 with about 17% of land being gentrified.

Results Based on City Scenarios

Though the trend lines are similar for most scenarios above, one of our contentions is that policies generally depend on specific clusters of variables in different contexts, but in the analyses just presented, we have averaged many different conditions. As demonstrated in Figs. 8.6, 8.7, and 8.8 below, our findings with the city case examples are largely similar to the generic cases above, with one exception.

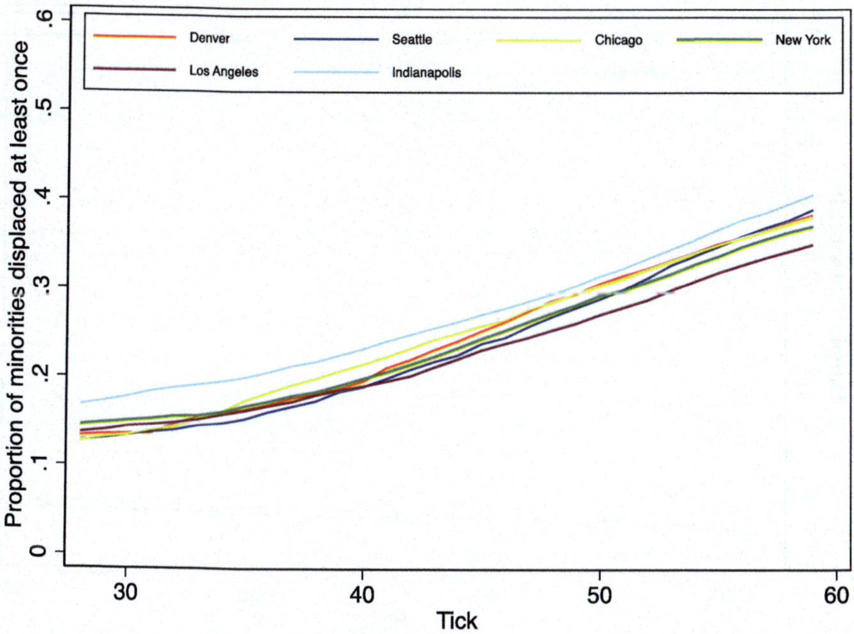

Fig. 8.6 Proportion of minorities displaced at least once

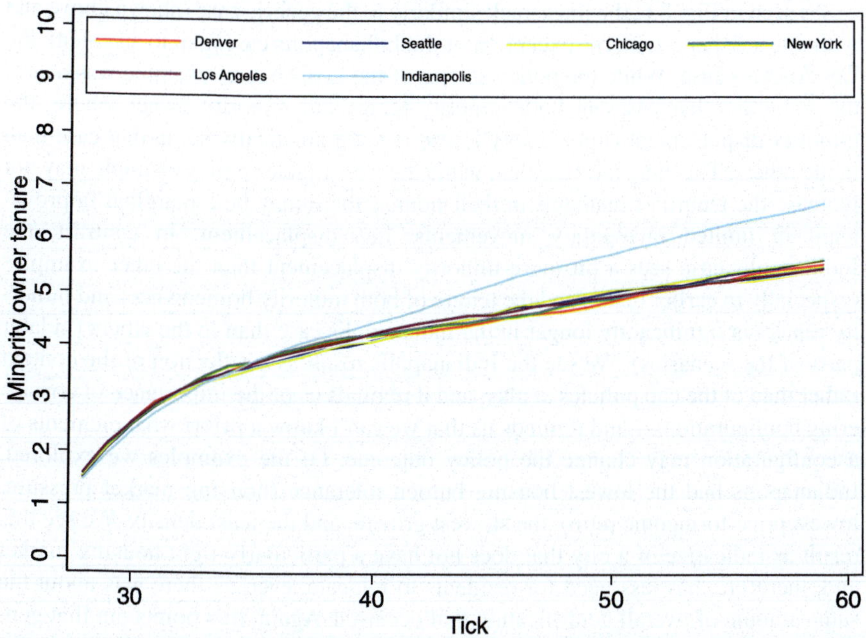

Fig. 8.7 Minority homeowner tenure

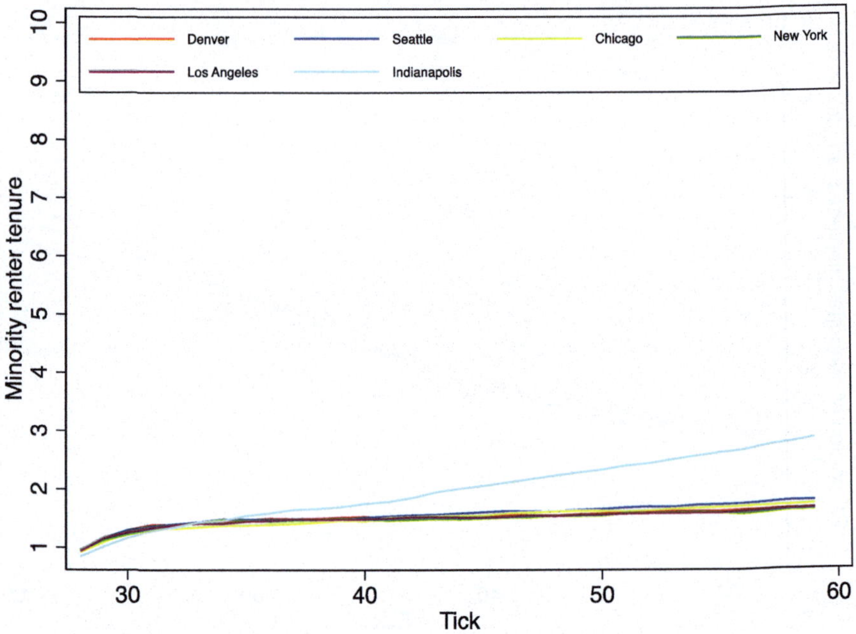

Fig. 8.8 Minority renter tenure

As seen in Fig. 8.6, the lines track similarly to the generic cases shown above and with one another—with the exception of the Indianapolis case and, to an extent, the Los Angeles case. While the policy setting of the Los Angeles example was nearly the same as other cases in both minority renter and minority owner tenure, the minority displacement is noticeably lower, if not dramatically so, in that case than in the others. This may indicate that, while the Los Angeles policy example may not increase the tenure of residents in their homes, there may be a marginal improvement in limited involuntary movements, i.e., displacement. In contrast, the Indianapolis case sees a bit more minority displacement than the other examples (especially in earlier ticks), and the tenure of both minority homeowners and minority renters is significantly longer in the Indianapolis case than in the others (in later parts of the scenarios). We see the Indianapolis result as a reflection of the context rather than of the cap policies at play, and it reminds us of the importance of considering configurations and reminds us that we can't know a priori what elements of a configuration may change the policy outcome. Of the examples we explored, Indianapolis had the lowest housing burden tolerance (housing market pressure, lowest price-to-income ratio), the slowest growth, and the least density. We see this result as indicative of a city that does not have a particularly tight housing market and, therefore, has less need for residents to relocate, even as there was about the same amount of overall gentrification that occurred. Again, this points out that gentrification and displacement are not the same process.

Discussion and Conclusion

In this chapter, we sought to explore how green gentrification and displacement played out in different settings, within the framework of existing policy environments. Specifically, we were interested in the effect that property tax caps and rent caps had on the extent and timing of gentrification-related displacement. However, we more broadly see these as indicative of the oft-recommended housing price controls. Comparing outcomes for renters and owners across a set of different contexts, we generally find that cap policies may have a relatively marginal effect on the displacement that occurs. These marginal effects occurred even with very tight price-increase limits of 2% for both owners and renters; as shown in Table 8.1, a 2% limit on price increases is the lowest of any US state at this time.

Across the different contexts that we utilized, one key finding is that cap policies seem to affect minority residents more than majority residents; majority resident outcomes were very consistent regardless of the policy scenario, whereas minority residential trends did change a bit. One reason for this might be our assumption that minority residents are poorer on average than majority residents. Other than for Asian Americans, this assumption is correct in the US. The actual impact of cap policies may be somewhat limited; i.e., if property values are growing at three or four percent a year (lower than empirical results for some areas of major cities), a cap on property taxes at 2% growth is a rather marginal difference that may not make that much real-world difference to wealthier residents. The effects of this marginal difference may only matter to the poor, for whom these marginal differences could constitute the difference between being displaced or not. The wealthy may not be particularly concerned about relatively small price differences, in which case housing market effects may drive more of the behavior in their choices than any marginal differences brought about by tax and rental caps.

Another reason may be the limitations of the similarity preferences that are inherent in the model. Minority residents will definitionally have fewer choices of plots that meet their similarity preferences, and, as the region begins to gentrify, they may find themselves forced out not just for financial reasons but rather because their existing residence no longer meets their similarity preference. In the model, we do not count residents as displaced when they relocate because their focal plot no longer meets their similarity preference (i.e., we only consider displacement on the basis of affordability), but leaving because of a similarity preference mismatch could be conceived of as cultural displacement. In our model, because minorities are *numerical* minorities (rather than only minoritized) this almost certainly affects minority residents more frequently.

We also find that virtually any cognizance of the potential for displacement offers a bit of respite from the worst effects of being displaced at least once brought about by our green gentrification scenarios. Across all the scenarios we tested, the scenario that did not consider gentrification at all (i.e., the scenario focused on getting the most environmental improvement possible) was consistently the worst in terms of displacement for minority residents. Whether the policy approach was more of a

JGE effort or included any combination of property tax caps and rent caps, we saw better outcomes for minority residents in particular. This suggests that—though none of the policies we analyzed is a silver bullet—any effort to consider and combat the risk of displacement is likely better than not considering displacement at all.

The lack of variability amongst these situations where gentrification was considered also points to the marginal effect that cap policies may have. Across all our trials, we saw that housing market effects tended to overwhelm efforts to limit displacement. This is perhaps not surprising, as property taxes and rent costs are just one of a host of considerations that people make when deciding where to live, both in the real world and in our model. In our model, the price of a plot is just one of five factors that agents consider in determining where to live, along with the environmental quality, proximity to amenities, density, and demographic makeup of the region around the plot. In the real world, similar factors, along with a host of other considerations like geographical aspects, school district considerations, and crime, all affect decisions about where to live. It is perhaps not surprising then that policies that affect one criterion, rate of increase of housing cost, seem to have relatively marginal implications for location decisions.

Nevertheless, this is not to suggest that price-focused policy efforts are not worth considering. Although the effects are marginal in our model, policies to cap the actual cost that people pay may indeed influence the displacement that occurs. They may not be magic bullets, but they may assist the most vulnerable in reducing displacement pressures a bit. We see this is as a positive implication of our results, but also a cautionary implication. Expecting property-tax and rent caps to eliminate or significantly reduce displacement seems overly sanguine. As a means of marginal protection for the most vulnerable, they may be helpful, but they are not likely to solve the problem of displacement. Understood in this way, cap policies may be a useful tool within a broader array of policies that, at the very least, consider the implications of gentrification on vulnerable populations.

References

Anguelovski, I., Connolly, J. J. T., Garcia-Lamarca, M., Cole, H., & Pearsall, H. (2019). New scholarly pathways on green gentrification: What does the urban 'green turn' mean and where is it going? *Progress in Human Geography, 43*(6), 1064–1086.

Anguelovski, I., Connolly, J. J. T., Cole, H., Garcia-Lamarca, M., Triguero-Mas, M., Baró, F., Martin, N., Conesa, D., Shokry, G., Pérez del Pulgar, C., Argüelles Ramos, L., Matheney, A., Gallez, E., Oscilowicz, E., López Máñez, J., Sarzo, B., Beltrán, M. A., & Martinez Minaya, J. (2022). Green gentrification in European and North American cities. *Nature Communications, 13*, 3816.

Black, K. J., & Richards, M. (2020). Eco-gentrification and who benefits from urban green amenities: NYC's high line. *Landscape and Urban Planning, 204*, 103900.

Bowman, J. H., Kenyon, D. A., Langley, A., & Paquin, B. P. (2009). *Property tax circuit breakers: Fair and cost-effective relief for taxpayers*. Lincoln Institute of Land Policy.

Dorazio, J. (2022, September 26). *Localized anti-displacement policies*. Center for American Progress. https://www.americanprogress.org/article/localized-anti-displacement-policies/

Immergluck, D., & Balan, T. (2018). Sustainable for whom? Green urban development, environmental gentrification, and the Atlanta Beltline. *Urban Geography, 39*(4), 546–562. https://doi.org/10.1080/02723638.2017.1360041

Jelks, N. O., Jennings, V., & Rigolon, A. (2021). Green gentrification and health: A scoping review. *International Journal of Environmental Research and Public Health, 18*(3), 907.

Lee, S. (2023, October). *Tax revolt and neighborhood change: Effect of assessment limits on neighborhood turnover*. Paper presented at the National Tax Association 116th Annual Conference on Taxation, Denver, CO.

Lyons, K., Farkas, S., & Johnson, N. (2007). *The property tax circuit breaker: An introduction and survey of current programs*. Center on Budget and Policy Priorities.

McConnell, B. (2019, March 19). *Learning from our mistakes: Anti-displacement strategies in Philadelphia*. NCRC. National Community Reinvestment Coalition. https://ncrc.org/gentrification-philadelphia/

National Apartment Association. (2023, May). Rent control—Statutory insights [PDF]. https://www.naahq.org/sites/default/files/2023-05/Rent%20Control%20-%20Statutory%20Insights.pdf

New York City Department of Finance. (2023). *NYC residential property taxes class one*. https://www.nyc.gov/assets/finance/downloads/pdf/brochures/class_1_guide.pdf

New York City Mayor's Office to Protect Tenants. (2024). *Fast facts about NYC housing*. NYC.gov. https://www.nyc.gov/content/tenantprotection/pages/fast-facts-about-housing-in-nyc

New York City Rent Guidelines Board. (2022). *Apartment and loft guidelines for leases commencing between October 1, 2022 and September 30, 2023* (Apartment & Loft Order #54). https://rentguidelinesboard.cityofnewyork.us/2022-23-apartment-loft-order-54/

New York City Rent Guidelines Board. (n.d.). *Housing types*. Retrieved December 18, 2023, from https://rentguidelinesboard.cityofnewyork.us/resources/apartment-hunting/housing-types/

Newman, K., & Wiley, E.K. (2006, July). Gentrification and displacement revisited: A fresh look at the New York City experience. *Research Bulletin, 31*. Centre for Urban and Community Studies, University of Toronto.

Quinton, J., Nesbitt, L., & Sax, D. (2022). How well do we know green gentrification? A systematic review of the methods. *Progress in Human Geography, 46*(4), 950–987.

Rigolon, A., & Christensen, J. (2019, December). Greening without gentrification. In *Parks and recreation*. National Recreation and Park Association. https://www.nrpa.org/parks-recreation-magazine/2019/december/greening-without-gentrification/

Rigolon, A., Stewart, W. P., & Gobster, P. H. (2020). What predicts the demand and sale of vacant public properties? Urban greening and gentrification in Chicago. *Cities, 107*, 102948.

Schilling, J., Tumber, C., & Velasco, G. (2023). *Greening America's smaller legacy cities*. Policy Focus Report. Lincoln Institute of Land Policy.

Wasi, N., & White, M. J. (2005). *Property tax limitations and mobility: The lock-in effect of California's proposition 13* (NBER Working Paper No. 11108).

Zuk, M., Bierbaum, A. H., Chapple, K., Gorska, K., Loukaitou-Sideris, A., Ong, P., & Thomas, T. (2015, August). *Gentrification, displacement and the role of public investment: A literature review*. Federal Reserve Bank of San Francisco, 79.

Chapter 9
Distillations and Policy Recommendations

> *We can therefore think about environmental justice as a dynamic process as well as an objective, a process of working towards environmental justice objectives and against forces which are serving to produce or sustain patterns of injustice.... But these will never finally resolve inequality and injustice always and forever, and in any case the terms in which these situations are understood will be dynamic rather than static and frozen in time (Walker, 2012, p. 221).*

Abstract Having reviewed the work of other scholars and performed our own analyses, here we distill findings and make policy recommendations. (1) Gentrification and displacement are not inevitably linked, and both are processes that take time, and even if the processes cannot be thwarted, they probably can be impacted. (2) Gentrification may be more situational than some seem to suggest, and displacement does not always follow, so in some cases concerns with green gentrification displacement are likely overblown, so avoiding cleaning and greening due to these fears may halt beneficial environmental improvements to no benefit. (3) Nonetheless, reducing green gentrification to zero is unlikely, and some cities are at greater risk and should be attentive to the risk of widespread gentrification-induced displacement, and some residents are at risk of multiple displacement, which is a concentrated hardship. (4) We review policies to address displacement, and note that they should be considered as part of a blend of approaches to affect not just financial implications, but cultural and geographic ones as well. (5) We end with a reminder that the wicked problems of Complex Adaptive Systems cannot be solved, but they can be managed.

Introduction

Considering cities as Complex Adaptive Systems (CAS) provides insight into why cities see so many different outcomes around cleaning, greening, gentrification, displacement, and overall environmental injustice. Trying to find one all-else-equal policy lever that performs the same across cities is usually futile because each CAS is different, and it is the configurations of situations, the individual structures of each CAS, that affect the likelihood of each outcome. In some ways, cities are

H. E. Campbell et al., *Green Gentrification and Environmental Injustice*,
https://doi.org/10.1007/978-3-031-65100-7_9

different from each other, and each city is dynamic in its own way such that changes make them different from themselves over time, and so efforts toward increasing environmental justice without triggering gentrification that leads to displacement cannot be solved once and for all, not for all cities, and likely not for even one city.

Still, we do not despair. Using the research of others and our own, we can identify different ways to understand these complexities of cities and also potential pathways that policy can influence and that can be effective parts of CAS. We hope that our work in this book can help local experts to find the present situation of their own city and derive some insights into how the specific nature of their city might affect how they manage concerns about green gentrification and displacement.

The next subsections detail key insights and policy recommendations we have distilled from the work presented in this book.

Insights

Insight 1: Green Gentrification and Displacement Are by No Means Certain

In Campbell et al. (2015), our policy advice for improving environmental justice was simple: Clean up brownfields, whenever and however you can in your city. But the green gentrification literature subsequently gave us pause: Had we provided advice that would create environmental gains but displace and harm the residents who had been suffering from local pollution? An insight from the analyses reviewed and created here is that the link may not be as clear, straightforward, and tight as existing green gentrification research has assumed—or feared. The amount of gentrification that occurs, and the extent of displacement that happens as a result of that gentrification, almost certainly varies in different contexts and it may not be as big a risk as policymakers fear.

We should all keep in mind the following points:

1. Cleaning and greening are beneficial for human health.
2. Gentrification is not an inevitable outcome of cleaning and/or greening (or other local investments).
3. Displacement is not an inevitable outcome of gentrification.
4. Even when gentrification and/or displacement occur, the process may unfold comparatively swiftly or slowly, depending upon contexts.
5. Neighborhoods can gentrify even *absent* cleaning and greening.

This last point is especially poignant: if we fail to clean and green in order to avoid gentrification and displacement, and yet they occur anyway, residents bear the costs without receiving any benefits.

The NCRC report finds, in "a comprehensive national level analysis of gentrification and displacement in 935 metropolitan areas" (Richardson et al., 2019, p. 8), that about 17% of urban tracts were eligible to gentrify, about 9% of gentrifiable tracts gentrified, and of those about 22% of tracts exhibited racial/ethnic displacement.[1]

In a study that is not intended to estimate the prevalence of gentrification, but to show the level of green gentrification in selected cities, Anguelovski et al. (2022) found that 17 of 28 cities studied in 9 countries provided evidence of green gentrification (60%), and "greening and greening alone" could cause gentrification in less than 50% of the 17 cities that experienced green gentrification. In the other 9 cities green interventions had an effect but only within the context of other factors.[2]

In our analyses, presented in Chaps. 5 through 8, always less than 60% of gentrifiable places gentrified, even with our fairly liberal operationalization of gentrification. All agent-based models (ABMs) are built with certain assumptions, and in our case we make assumptions about the type of cities under consideration (e.g., growing cities, with specific segregatory preferences leading to certain levels of segregation, and with density preferences and levels). Thus, the percent values we provide here and below must be understood keeping in mind the setting of the simulation model, rather than taking them as being equivalent to empirical estimates. ABM outputs are most useful for comparing in a relative sense.

Still, in Chap. 5, Fig. 5.2, between 25% and 40% of land is gentrifiable, but at most only about 30% of all land ever gentrifies even after brownfield cleanup interventions. In Chap. 6, looking at empirical data for 26 gentrifiable Chicago census tracts, 10 of which received a new park during a period when Chicago was known to exhibit very high rates of gentrification, about 58% of the tracts gentrified, including some that did *not* receive new parks. In ABM analysis in Chap. 7, even after green interventions, at most 40% of land gentrified (Fig. 7.1). In Chap. 8, where we introduce policies specifically intended to limit the extent of gentrification and related displacement, less than 20% of land gentrifies.

In the ABM analysis in Chap. 5, we do see some cases where gentrification *displacement* eventually, over more than 30 ticks (which we can think of as about 30 years), rises to about 80%—though in the first decade or so it does not rise above 40%. In the ABM analysis in Chap. 7 displacement never rises above about 20%, and in Chap. 8 minority displacement never rises above 50%.[3] This is not to imply that there is no real harm in this level of displacement, but rather to note that, in the aggregate, there are many factors that affect the extent of displacement that occurs, and displacement that takes place over a span of many years is fundamentally different from that which takes place in a matter of months.

[1] See Richardson et al. (2019), or Chap. 1 herein, especially Fig. 1.3, for their criteria and study period. They note that about 67,153 census tracts are defined as urban in their study period, so about 11,200 tracts were eligible to gentrify and about 1000 did gentrify as defined (p. 13).

[2] See Anguelovski et al. (2022), subsection entitled "Greening versus other growth-oriented drivers of gentrification."

[3] In comparing various figures in this subsection, it is important to note that how many *places*, such as census tracts or cities, saw significant displacement, and how many *people* were displaced are not the same.

Gentrification and displacement are time-based processes, and the duration of change matters to humans. The longer that people can stay in a community where the environment is improving, the better off they are, certainly for health purposes, but also for their sense of community and avoiding the burden of moving costs and relocation to areas with (potentially) worse environmental conditions. If we can prolong the time it takes for displacement to occur, we enable people to reap benefits from a better environment; build wealth in their homes (provided they are homeowners); potentially improve their social capital (via social mix); and to decide if, when, and how they would like to leave voluntarily.

At some point over the duration of time, displacement becomes something different. We know that neighborhoods always change over time, but the timing of gentrification and displacement have rarely been examined or elaborated upon. There is a sense that gentrification is something that occurs in a short time frame, but that time frame is not well specified. We might all agree that a rapid demographic change over a period of a year or two would be classified as gentrification and that anyone who is forced out because of rising housing costs is displaced, but what if the process takes a decade? A generation? All neighborhoods change to some extent in the span of a decade, so at what point is gentrification different from normal change that just naturally occurs as residents move in and out of a community?

Lesson 1

Our lesson from this is that a generalized *fear* of harm caused by quick gentrification and displacement should not prevent the *almost-certain* goods created by cleaning and/or greening. Instead, we should think carefully about how we can reduce the amount of displacement, lengthen the time to displacement, and reduce the harm to those likely to be displaced.

Insight 2: Some Cities, Places, and Groups Do Need to Worry About Displacement More Than Others, so They Need to Think More Carefully About How to Proceed

Nonetheless, in some sets of circumstances, the likelihood of quickly occurring gentrification being triggered by environmental improvement—cleaning or greening—is high, and the likelihood in some circumstances that gentrification will lead to displacement is high. Based on our review of the empirical literature and our own findings here, we expect that the risk of gentrification and displacement will be higher when at least some of the following circumstances are present, and some clusters of these may make it especially high.

1. Larger urban regions, particularly those with metropolitan populations over 1,000,000, seem more prone to gentrification (although not necessarily more prone to displacement).
2. When density is already high, then improvements seem more likely to lead to gentrification and also to related displacement.
3. Citywide programs may not be the best approach, as areas near Central Business Districts (CBDs) may be more prone to gentrify.[4]
4. Gentrification seems to follow a spatial flow, as neighborhoods adjacent to gentrifying neighborhoods are more likely to gentrify—even without any environmental improvements.
5. Neighborhoods that have a set of development improvements in play—already several improvements are in action and cleaning/greening is added to the mix—are more likely to gentrify.[5]
6. Neighborhoods that are nature-poor may be more likely to gentrify with a green intervention,[6] but the cluster of conditions matters.[7]
7. There may be more gentrification and also more displacement in regions where there are *both* high density and low segregatory preferences.

Richardson et al. (2019) find that gentrification is much more likely in US Core Based Statistical Areas (CBSAs) with populations of 1,000,000 or more; CBSAs with 1,000,000 or more people contained 79% of gentrifying tracts even though these cities included only about half of the total US population.

Further, both Richardson et al. (2019) and Anguelovski et al. (2022) find that tracts near Central Business Districts (CBDs) are more likely to gentrify, and density matters. Areas that are already dense lack the protective buffers available to areas with less density, which can accept more in-comers with less gentrification or slower gentrification, and can accommodate more gentrification with less or slower displacement.

Anguelovski and colleagues (2022) find that green gentrification needs to be considered within the broader array of other local policy interventions and regional sustainability initiatives. This suggests that adding a green intervention to a neighborhood that is already undergoing other initiatives, including new building, is more likely to increase gentrification risk. This finding also supports the importance of considering configurations—several variables at once—rather than just one at a time, and is supported by the spillover effects observed in Chap. 6.

Chapter 5 indicates that gentrification is more likely when density is higher, but also indicates that this interacts with segregation such that areas with higher density *and* lower segregatory preferences are more readily gentrified. Once gentrification occurs, we find *much more* minority displacement when there is *less* segregation

---

[4] See Eckerd (2011).

[5] This insight again emphasizes the importance of context.

[6] This may be because they exhibit larger rent gaps than other areas. See Diappi and Bolchi (2008).

[7] Compare results in Chaps. 6 (where what can be thought of as park deserts gentrify) and 7 (where some combinations that include locating parks in park-desert areas perform well).

interacted with high density (Figs. 5.2 and 5.3). Chapter 6 indicates that, even without a green intervention, gentrification is more likely when nearby tracts are already gentrifying—so there is a risk of gentrification spillover. Chapter 6 also indicates that *green* gentrification is more likely when an area is nature-poor before the green intervention, and this finding accords with Wolch et al. (2014), who find any greening can cause gentrification if there was little greenspace and high density and high urban heat.

Lesson 2

It is unlikely that any one policy approach will provide a magic bullet against displacement; gentrification and displacement are parts of a complex process within a CAS, and policies to manage these problems likely require a suite of policy approaches—a policy mix tailored to each city's own contexts rather than just one.[8] Searching for the *single best policy* to prevent gentrification and/or displacement may not work in settings where different clusters of local conditions occur and influence the effectiveness of intended actions. Instead of seeking single best policies, or even trying to avoid displacement, we suggest a more realistic approach: recognize the unique cluster of conditions present in a region and think about a suite of approaches that are focused on minimizing the amount of and the harm attributed to the displacement that happens.

Insight 3: Those Who Are Displaced Once May Be Displaced Multiple Times

When developing our model, we began tabulating the number of times that a resident was displaced, not necessarily for analytical purposes, but rather for testing the model. However, as we tested the model and ran trials, we noted that residents who were displaced once tended to be displaced more than once over the course of a trial run. We ultimately found that those who are displaced once may be displaced multiple times. In our model, we see that displacement is a common occurrence for those minority residents displaced at least once. If a minority resident is displaced at least once, then they were likely displaced at least a second, and perhaps a third time over the entire course of the model's time period.

From an analytical perspective, this makes sense. The poor are predominantly affected by displacement, and if they are displaced from an area because the price is too high, then it stands to reason that they are displaced to an area with lower

[8] The term "policy mix" captures the idea of the reality of using multiple policy tools to address complex social problems in the field of environmental policy, climate-change policy, or innovation policy (Howlett, 2019).

prices—the very sort of area that is definitionally also at risk of gentrification. In fact, during our model runs, we found that about half the time that residents were displaced from gentrifying areas, they tended to relocate to gentrifiable areas, which obviously put them at risk of being displaced a second time (and then a third time and so on).

Although one addendum to the NCRC report highlighted above suggests this possibility (Mitchell, 2019), the gentrification literature mostly does not. In the literature, gentrification is typically treated as a characteristic of a neighborhood, as is displacement to an extent. When treated as a characteristic of neighborhoods, it is considered as a rate (what percent are displaced), but displacement can also be treated as a variable measured at the individual or family level, and, when it is, it is typically treated dichotomously (i.e., a resident is displaced, or not, from just the focal neighborhood). There is no doubt that a true real-life challenge here is tracking the movements of people after they have been displaced. It would be very difficult to actually observe an individual or family being displaced a second or third time—let alone the challenge of even observing them being displaced the first time.

Nevertheless, our results suggest that, while the literature as a whole might be overstating the risk of displacement after the placement of an environmental amenity, it might also be understating the true burden of displacement for those who are displaced. It seems that, most of the time, most people are not at risk of displacement. At the same time, however, it seems that, once someone is displaced, the risk of them being displaced a second time increases. This is a true hardship, especially for the poor—and our simulation results indicate it is the minority poor most at risk of this.

At the same time, we need to consider this result in the context of our recognition that the timing of gentrification and of displacement matters. Being displaced unexpectedly is obviously a burden, but being displaced more than once in a relatively short time span is a substantial burden.[9] In our simulation trials, we had scenarios where some residents were quickly displaced and also scenarios where important numbers of displaced residents experienced displacement multiple times. In some cases this was in the same scenario but, under other sets of circumstances, we saw different manifestations of displacement. We therefore urge consideration of timing: The experience of being displaced twice in 2 years is not the same as the experience of being displaced twice in two decades—or twice in 30 years.

One broad finding is that displacement is perhaps much more nuanced that it has generally been portrayed to be, encompassing aspects of timing, repetition, and the circumstances that people find themselves in after they have been displaced. Each of these considerations may require a different approach to managing the displacement problem.

[9] Mitchell reports that she and her husband were involuntarily displaced 4 times in 5 years (Mitchell, 2019); anyone who has ever moved knows that is a burden.

Lesson 3

Our lesson from this is that "Displacement" isn't just one problem. Green displacement—or any gentrification displacement—can be a burden for long-time residents, but the pace and frequency of displacement, and where residents are displaced to, are also relevant problem parameters.

Local experts need to consider the nature of the problem in *their* city and/or relevant neighborhoods. If the same poor residents are displaced again and again, that is a worse problem; if people are displaced to worse neighborhoods, that is a worse problem. Decision-makers should stop thinking in terms of stopping displacement—for it likely cannot be stopped—but can start thinking specifically in terms of reducing the costs for those who are displaced. It may be possible to reduce the rate of displacement, and it should be possible to help people with the problem of displacement. For example, cities could provide relocation services: A poor person who has just been notified that they can't pay their new rent and they will have to move is unlikely to have much bandwidth to figure out their best new location; city services could help.

Insight 4: Smart Policy Interventions Are Still Possible

Based on local conditions and clusters of variables, we believe that the research presented here suggests some useful policy interventions.

1. Clean up brownfields to improve health and environmental justice. This is unlikely to cause much displacement outside of high-density, low-segregatory-preference cites, and the latter seem to be relatively rare in the US at this time, as indicated by segregation levels in high-density cities.[10]
2. Don't insouciantly also green where brownfield cleanup has occurred; this is likely to cause displacement. If doing this, be sure to consider displacement-amelioration strategies.
3. Consider just-green-enough (JGE) strategies: add greening to the lowest-priced areas, but without extra amenities that are likely to attract incomers. The greening component should improve health, generally of the poorest. It is also likely to generate gentrification and displacement, but relatively slowly. With a longer period of time before displacement, residents are allowed a longer tenure before they must move. Within this context, consider placing fewer larger greenspaces, rather than more smaller ones.

[10] Again, we note that in a CAS observed levels of an outcomes such as segregation are not the same as an input such as segregatory preferences, but quite a bit of evidence suggests that these two are positively correlated.

4. Add high-desirability greening to park-desert areas; but recognize that when there is nearby gentrification or other clusters of neighborhood-improving investment then gentrification and displacement are more likely.
5. If you are truly concerned about green gentrification and displacement, then steps to reduce the pace of housing-price increase (such as via tax caps or rent caps) can make some beneficial difference—if these are legal in your state.

Just cleaning up brownfields can be expected to have health benefits and does not create serious extreme gentrification and displacement pressures—though such pressures are stronger in cities with high density and low segregatory preferences (perhaps such as Seattle, which has low-to-medium segregation and population density of about 8800 persons per square mile).

As indicated in our ABMs, the JGE scenarios (such as small parks in low-priced areas, with low-desirability added amenities) may generally see more displacement, but that displacement takes place over a longer period of time. This suggests that smaller, less desirable parks foster more displacements, but at the same time offer the residents who will be displaced longer tenure in the improved neighborhood before they are displaced. Displacement is slowed.

Unfortunately (though in keeping with JGE), an approach that focuses the most directly on environmental improvement by adding lots of high-desirability small parks near pollution (i.e., with amenities that extend beyond the environmental improvement that parks provide) leads to the most common experience of displacement. Still, in many scenarios, the proportion of minorities displaced at least once in the first 10 years is low, which suggests that at least sometimes cleaning and greening is also reasonably sustainable for local residents. Also, the scenarios suggest that placing high desirability parks in park-desert areas (e.g., locating a new park far from other greenspace) generates the least minority displacement of all, another hopeful finding.

Chapter 7 specifically addresses green gentrification and displacement and seeks to compare different approaches to green intervention, including interventions that may be thought of as JGE—namely, small, low-desirability parks on low-priced land—and what might be thought of as sister to our original idea of cleaning up brownfields—namely, choosing to place parks near pollution. The results indicate that the JGE scenario should reduce gentrification risk compared to the single-eyed focus on environmental-improvement.

Lesson 4

Clean up brownfields to improve health and well-being. Provide greening to improve health and well-being, but be sure to consider effects on displacement and how they may be ameliorated.

Insight 5: What Factors Don't Seem to Matter?

Our research also indicates that some policies or decision choices may not have much effect on green gentrification and displacement, as described here.

1. Alone, retaining some type of undesirable land use within a greened area should not be expected to reduce green gentrification. For example, industrial zoning, or the presence of government-provided affordable housing (e.g., public housing)—which is viewed as undesirable in some literature— still allows gentrification in the right situations.[11]
2. Using blanket policies, such as targeting cleanup to the poorest communities or to the most polluted places, is not particularly effective. We tended to find that the specific approach was less important than that the cleaning took place to begin with. Our model suggests there is no need to be rigid in determining which areas to improve.
3. Focusing on the size characteristics of greenspaces that are provided may just be a small part of the bigger picture. Large or small parks aren't truly the key; it's the *combination* of size and other configural elements, such as suppressive forces like government housing, lack of pre-existing greenspace, etc.
4. Tax caps and/or rent caps are unlikely to prevent displacement. Having any one of these caps seems to have the same effect as any other (alone or in combination) and the displacement-reduction effects of these policies seem to be marginal; they provide only a small benefit. They can be expected slightly to reduce displacement, but not substantially. However, what benefits there are accrue to minority residents.
5. From Chap. 5, we learn that whether to clean *and* green the same space, or just to clean a brownfield without developing it, may not be important in terms of avoiding displacement.

Lesson 5

Neither the details about cleaning up hazardous sites, nor the particulars about placing new greenspaces, seem to have much of an impact on the extent of displacement that occurs. Minimizing displacement and its harms is about working with the people that are at risk of displacement. Land-use and tax policies have a place within a set of policies intended to minimize displacement, but they appear to be supplemental tools rather than primary tools.

[11] See Gould and Lewis (2017); and Chap. 6 in this book.

Insight 6: Directions for Future Research

In addition to these policy recommendations, we see several implications for future research, most of which are related to the policy recommendations above.

1. *Conceptually distinguish between gentrification and displacement*

   As we have noted several times, gentrification and displacement are related, but conceptually distinct, processes. The literature has begun to recognize this important distinction, but not often in a way that acknowledges that the two processes can be, and often are, decoupled. Of course they are also often coupled, but it really depends upon the context. In certain settings, it appears that gentrification can be a policy goal with minimal concern about displacement. For better analysis, these two concepts should be kept distinct.
2. *Add nuance to the concept of displacement*

   One of our key conclusions is that the nature of displacement is more nuanced than it has often been portrayed. Displacement has often been framed as a dichotomous outcome—either a resident is displaced or not. However, the characteristics of the displacement that happens are important too. Displacement that happens quickly and unexpectedly is almost certainly more burdensome than displacement that plays out over a longer time period. Being displaced multiple times adds to the burden a resident experiences, likely substantially, and it may happen repeatedly for the most vulnerable residential populations. And the support that residents receive, while not eliminating the burden of displacement, may help mitigate it. Though we fully recognize that this is an empirically tough task, we believe there needs to be more research on the actual displacement experiences of residents—how quickly, to where, and how often?
3. *Remember that policy efforts in CAS are contingent on context, aka configurations or recipes*

   We suggest that scholars need to recognize that the effects of policy approaches applied to systems that are importantly complex and adaptive are contingent on the context and the unique clusters of conditions that happen to exist. This means that a policy approach that works in one context may not work in another. Specific to this research, it also suggests that any policy that improves conditions for those at risk of displacement is also contingent on timing. Just because a policy works today does not mean that it will continue working indefinitely, and, importantly in our federal system, some policies are not allowed in all contexts. For example, while rent control policies may offer some marginal benefits to those at risk of displacement, most states do not permit local governments to put these programs in place.

Conclusions

Cities' environmental conditions can and will change over time. This has at least two implications: We may be able to identify a best policy at one point but can't expect it will be best forever. *When* and *where* a policy should be introduced may be as critical a question as *which* policy should be crafted. The city is a complex *adaptive* system. It is the nature of a CAS that we cannot predict exactly, and that clusters of conditions matter. It is the nature of "the problem of organized complexity" (Weaver, 1948) and also the nature of "wicked problems" (Rittel & Webber, 1973) when a significant number of elements are interrelated into an organic system whole.

Nonetheless, the fact that policymakers cannot predict as well as they can in more linear or deterministic systems does not mean that there are no tools to explore the clusters of conditions that can help us understand how a policy will play out. As we illustrate herein, there are CAS-appropriate tools such as QCA and ABM. Most of all, policymakers and practitioners need to keep an eye on changes to the clusters of key conditions that matter in their own neighborhoods.

Cities as CAS are self-organizing systems within which a myriad of human and non-human actors interact and create dynamic environments. It is difficult to control such systems because these actors can adapt or adjust to changing environments without a full understanding of their own actions and consequences. We cannot predict exactly how such systems would unfold with certainty. Note that Florida's (2003) "creative class" approach did often work in revitalizing areas, but also produced unintended consequences of gentrification and displacement. Still, we may be able to understand the range of the possible behavior of such systems, at least in the short run, if we can identify relevant clusters of local conditions to examine and with which to experiment. City decision-makers can look at the clusters in their communities and then have better (but not perfect) abilities to anticipate what is likely to happen. The insights from CAS-sensitive approaches help, but are not a silver bullet.

Given the nature of the city as CAS, a policy approach that focuses on prolonging the time it takes until displacement happens seems more likely to reduce the harm of displacement than a policy effort intended to stop displacement from occurring. We do not suggest this because we think that displacement is a good outcome. We suggest it because some level of displacement seems inevitable under most conditions. Our research suggests that housing-market forces are stronger than policy, and as long as demand for housing is increasing in a community, prices will rise and people will be displaced.

What we believe can be done, however, is to use policy and administration to lessen the harm caused by displacement. This may mean working with communities to identify people or properties most at risk of displacement and directly intervening to offer support in case of displacement, such as offering home-search services, or working with property companies to negotiate more favorable rents for the displaced or those at risk of displacement. It may mean offering direct financial support to allow people to avoid displacement, at least for a time. Under the right set of circumstances, it may mean working with communities on JGE strategies to try to lengthen the time before displacement.

Restatement of Some Key Considerations

1. Against our expectations, the park policies leading to the least gentrification and displacement risks involve high-desirability parks, placed on either low-priced land or in park-desert areas.
   (a) Within the ABM, these strategies are estimated to lead to minority displacement of less than 10%
   (b) But policymakers should be aware! Those residents who are displaced once are at risk of being displaced again, which can be a true hardship for them.
2. A JGE approach does not necessarily stop either gentrification or displacement, but it might make the process play out over a longer time frame, which, while maybe not ideal, is probably better for most residents than a process that plays out rapidly.
3. It may also be the case that any consideration of the risk of gentrification and displacement will be better than ignoring the interdependent nature of environmental improvement and gentrification/displacement. While we did not find that any policy approach was particularly effective at stopping displacement from occurring, we did find that being aware that gentrification could result from environmental improvements, and doing almost anything to ameliorate the risk of displacement, was better than doing nothing about it at all. The bad news on this front was that no policy approach seemed to have anything more than a marginal effect, but the good news is that almost any policy approach seems likely to make things at least a bit better.
   (a) In particular, the housing-cost reduction policies analyzed in Chap. 8 are estimated to make things a bit better, with almost no impact for majorities, but some benefits for minorities.
4. Because policies designed to control housing costs (e.g., property tax and rent regulation) are governed by state and local governments, it is important to be aware of various existing policy options and constraints. Property tax caps and rent caps seem to have a small impact in general, but the impact for minorities seems to be bigger. Policymakers and planners may need to think about including public communication campaigns including community engagement in their projects to provide options for residents to prevent immediate displacement.

The worst burdens of displacement seem likely to happen when the displacement is unexpected, when it happens quickly, or when it happens to someone more than once in a short period of time. Therefore, our policy recommendations are focused on four key goals:

1. Disaggregating the concepts of gentrification and displacement and recognizing that gentrification will not always lead to high levels of displacement, therefore suggesting that avoiding necessary environmental improvements for fear of gentrification and displacement might be making things worse for residents by not providing environmental improvements that could genuinely improve peoples' lives.
2. Shifting the focus away from avoiding displacement to a focus on limiting it, and ameliorating the worst burdens associated with it.
3. Paying attention to the timing of both the gentrification process and any subsequent displacement that occurs, with a recognition that, all else equal, gentrification that takes longer to play out probably means that any displacement that occurs will also take longer to play out, which is a better outcome than rapid gentrification and displacement.
4. Choosing feasible policy options within a given jurisdiction. Because policies designed to control housing costs (e.g., property tax and rent regulation) are governed by state and local governments, it is important to be aware of various policy options and constraints. Property tax caps and rent caps are hopeful for minorities, but not allowed everywhere.

We conclude the book with two main points. First, we have summarized a new frame for analyzing social problems having to do with different configurations. In order to avoid unintended consequences of urban policy, people need to fully assess the key clusters of interrelated conditions before they start thinking about decisions. Second, we provide a set of policy recommendations that are based upon our analyses and the clusters that different cities may face, drawing from real-world examples of cities that fit these different cluster types. Taking the complexity of urban systems seriously reveals that just because a policy intervention works in one place does not mean it will work in another. Therefore, we do not identify best practices—we will not suggest applying a policy that seems to work in one place to some other setting. We suggest policies that may improve or exacerbate conditions within different types of settings. We argue for using a critical perspective for understanding a problem within its context, and for shifting a focus towards managing complexity rather than simplifying it.

References

Anguelovski, I., Connolly, J. J. T., Cole, H., Garcia-Lamarca, M., Triguero-Mas, M., Baró, F., Martin, N., Conesa, D., Shokry, G., Del Pulgar, C. P., Ramos, L. A., Matheney, A., Gallez, E., Oscilowicz, E., Máñez, J. L., Sarzo, B., Beltrán, M. A., & Minaya, J. M. (2022). Green gentrification in European and North American cities. *Nature Communications, 13*(1), 3816. https://doi.org/10.1038/s41467-022-31572-1

Campbell, H. E., Kim, Y., & Eckerd, A. (2015). *Rethinking environmental justice in sustainable cities: Insights from agent-based modeling*. Routledge.

Diappi, L., & Bolchi, P. (2008). Smith's rent gap theory and local real estate dynamics: A multi-agent model. *Computers, Environment and Urban Systems, 32*(1), 6–18.

Eckerd, A. (2011). Cleaning up without clearing out? A spatial assessment of environmental gentrification. *Urban Affairs Review, 47*(1), 31–59.
Florida, R. (2003). *Rise of the creative class*. Basic Books.
Gould, K. A., & Lewis, T. L. (2017). *Green gentrification: Urban sustainability and the struggle for environmental justice*. Routledge.
Howlett, M. (2019). *Designing public policies: Principles and instruments* (2nd ed.). Routledge.
Mitchell, S. (2019, March 18). *In Richmond, Virginia, gentrification is colonization*. NCRC. https://ncrc.org/gentrification-richmondva/
Richardson, J., Mitchell, B., & Franco, J. (2019, March 19). *Shifting neighborhoods: Gentrification and cultural displacement in American cities*. NCRC: National Community Reinvestment Coalition. Accessed at https://ncrc.org/gentrification/
Rittel, H. W. J., & Webber, M. M. (1973). Dilemmas in a general theory of planning. *Policy Sciences, 4*, 155–169. https://doi.org/10.1007/BF01405730
Walker, G. (2012). *Environmental justice: Concepts, evidence and politics*. Routledge.
Weaver, W. (1948). Science and complexity. *American Scientist, 36*(4), 536–544.
Wolch, J. R., Byrne, J., & Newell, J. P. (2014). Urban green space, public health, and environmental justice: The challenge of making cities 'just green enough'. *Landscape and Urban Planning, 125*, 234–244.

Index

A

ABM, *see* Agent-based model/agent-based modeling (ABM)
Affordable housing, 4, 10–13, 20, 23, 34, 63, 106, 108, 116, 117, 119, 122, 163, 192
African Americans, 9, 13, 23–25, 28, 30, 75, 81, 115, 153
  See also Black
Agent-based model/agent-based modeling (ABM), 2, 51, 54–56, 74, 77, 78, 93, 96, 97, 132, 134–136, 138, 139, 151–154, 158, 167–173, 185, 191, 194, 195
Air pollution, 24
Albuquerque, 8
Amenities, 1–4, 9, 11–15, 25, 26, 30–32, 34, 37, 47, 48, 62, 65, 67, 71, 73–97, 114, 122, 132, 133, 135–137, 139, 150–153, 165, 180, 189–191
Asians, 7, 24, 30, 179
Assessment limits, 160–164, 166
Atlanta, 8, 23, 32
  Beltline, 32, 166
Austin, 30

B

Baltimore, 3, 6, 8
Black, 7–9, 20, 21, 29, 30, 32, 115
  See also African Americans
Blight, 5, 12–14, 33, 34, 48
Boston, 10, 30
Brooklyn, 30, 32, 35
Brownfields, 1, 19–21, 24, 30–33, 54, 116, 184, 185, 190–192
Bullard, L.M., 23
Bullard, R.D., 23

C

CalEnviroScreen, 25
Cascade, 34
Case studies, 2, 35, 38, 51, 57, 78, 109, 111, 113–117
CBA, *see* Cost-benefit analysis (CBA)
Census, 7, 12, 23, 29, 30, 80, 81, 98, 113–117, 126, 132, 139, 153, 164, 168, 173, 185
Charleston, 95
Charlotte, 163, 164
Chennai, 35
Chicago, 3, 6, 8, 35, 46, 48, 95, 106, 108, 109, 111, 113–117, 120, 122, 131, 133, 160, 164–166, 172, 185
Chicago neighborhoods
  Bridgeport, 124
  Logan Square, 119, 120, 123
  Near West Side, 117, 123, 124
  South Lawndale, 123, 124
  West Town, 123, 124
Chicago's Environmental Action Agenda, 113
China, 32
Circuit-breakers, 159–166, 168
Cleaning and greening, 2, 3, 23, 30, 31, 33–38, 66, 106, 131, 132, 139, 140, 142, 158, 184, 191

H. E. Campbell et al., *Green Gentrification and Environmental Injustice*, https://doi.org/10.1007/978-3-031-65100-7

Clusters, 38, 52, 54, 86, 105–127, 143, 171, 176, 186–188, 190, 191, 193, 194, 196
Columbus, 163, 164
Community preferences, 6
Complexity
approach, 43–57
theory, 38
Complex systems, 2, 37, 51, 77
Configural, 2, 15, 33, 38, 57, 192
Configurations, 2, 11, 16, 51–54, 110, 112, 118, 119, 125, 140, 150, 171, 178, 183, 187, 193, 196
Conjunctional causes, 53, 118
Cost-benefit analysis (CBA), 37
Criteria pollutants, 24
Cultural displacement, 7, 15, 114, 179
Curitiba, 33

D
DC, *see* Washington, DC
Demand, 9–11, 15, 26, 62, 63, 65–67, 84, 131, 145, 194
Densification, 68
Density, 12, 21, 31, 32, 34, 49, 64, 73–98, 133, 135, 149, 152, 153, 172, 173, 178, 180, 185, 187, 188, 190, 191
Denver, 8, 30, 163–166, 171, 172
Disamenities, 1, 11, 15, 30, 34, 76, 78–80, 82–88, 94, 95, 97, 115, 135, 137, 138, 147, 151–153, 158
Displacement, 1, 20, 43, 61, 107, 132, 157, 183
Durham, 95

E
Eco gentrification, *see* Green gentrification
Economic development, 2, 5, 6, 11, 26, 27, 29, 50, 63, 69, 73, 75, 76, 88, 93, 120, 121
Economic rents, 63
Economic spillover, 106, 108, 109, 116, 117, 119, 121
Education, 48, 166
EJ, *see* Environmental justice (EJ)
11th Street Bridge Park, 166
Environmental bads, 22, 25
Environmental goods, 22, 25, 131
Environmental improvements, 6, 26, 32, 35, 43, 50, 66, 68, 69, 71, 93, 97, 131–133, 136–139, 142, 147, 148, 150, 157–159, 166–168, 179, 186, 187, 191, 195, 196
Environmental injustice, 1, 6, 15, 19–38, 45, 49, 51, 55, 90, 121, 139, 140, 158, 173, 183
Environmental justice (EJ), 2, 3, 7, 19–26, 29, 30, 32, 33, 35, 36, 43, 48, 49, 53, 55, 64, 65, 68, 77, 81, 95, 99, 114, 120, 132, 139, 147, 167, 184, 190
Environmental justice communities, 32
Ethnicity, 22, 25, 36, 80, 90, 94, 139, 140
Externality
negative, 31
positive, 31

G
Gentrifiable, 7, 8, 13, 29, 62, 64, 65, 85, 86, 89, 90, 121, 134, 135, 145, 168, 170, 185, 189
Gentrification, 2, 20, 43, 61, 73, 106, 131, 157, 183
Gentrified, 3, 6–8, 13, 27, 28, 30, 62, 64, 65, 67, 68, 71, 78, 85, 86, 89, 90, 134, 135, 139–143, 145, 149, 169, 176, 185, 187
Gentrifiers, 10, 26, 32, 34–36, 63, 65, 94, 115, 137
Grand Rapids, 31
Green gentrification, 1–16, 20–22, 26, 30–34, 36, 37, 51, 61, 105–122, 125, 131–134, 136, 157–180, 184–188, 191, 192
Green growth machine, 118, 119, 121
Green-growth-machine politics, 31
Green investments, 22, 107, 108, 122
Greenspaces, 9, 14, 20, 22, 24, 25, 31, 32, 35–37, 106–109, 112–116, 119–121, 131, 158, 188, 190–192
Greenup, 36, 121
Growth machine, 31
Growth-machine politics, 31

H
High Line Park, 158, 159, 166
Hispanics, 7–9, 24, 25, 29, 30, 75, 115, 126
Homeowners, 14, 21, 63, 81, 158–171, 174, 175, 177, 178, 186
Housing tenure, 160, 163
Houston, 10, 23, 30, 160, 164

I
Incomers, 4, 9, 13, 36, 97, 137, 150, 187, 190
Incomes, 3, 5, 7, 10, 11, 13–15, 21–24, 27–29, 31–36, 53, 61, 63, 66, 68, 70, 71, 80, 81, 94, 106–108, 114–116, 120, 126, 139, 153, 159–166, 173
Indianapolis, 48, 95, 163, 164, 172, 178
Invisible hand, 36, 37

J
Jacksonville, 164
JGE, *see* Just green enough (JGE)
Just green enough (JGE), 14, 26, 34–37, 68, 107, 133, 137, 139, 142, 144–146, 148–151, 168, 171–173, 175, 176, 180, 190, 191, 195

K
Kaldor-Hicks principle, 37

L
Land constraints, 10
Land trusts, 34, 158, 166
Latinos, *see* Hispanics
Latinx, *see* Hispanics
Linear green space, 36
London, 27, 61, 133
Los Angeles, 3, 8, 35, 158, 160, 164–166, 172, 178
Los Angeles' River Revitalization, 166

M
Millennial generation, 3
Milwaukee, 48
Multiple pathways, 117
Multi-value Qualitative comparative analysis (mv-QCA), 113, 115, 116

N
Nantes, 107
National Community Reinvestment Coalition (NCRC), 3, 6, 7, 13, 29, 34, 75, 163, 184, 189
National Priority List (NPL), *see* Superfund
NCRC, *see* National Community Reinvestment Coalition (NCRC)
Neighborhood changes, 3, 5, 14, 15, 21, 31, 33, 36, 46–48, 62, 66, 69, 74, 106, 114, 186
New Orleans, 8
New York City (NYC), 3, 6, 8, 32, 35, 36, 64, 83, 94, 133, 158–160, 162–166, 172
Noise pollution, 24
Non-market approaches, 33
Normalized Difference Vegetation Index (NDVI), 115
NPL, *see* National Priority List (NPL)
NYC, *see* New York City (NYC)

O
Oklahoma, 164
Omaha, 95
Ostrom, E., 2, 48
Owners, *see* Homeowners

P
Pareto efficiency, 37
Park
  park deserts, 138, 141, 142, 187, 191, 195
  park distribution, 136
  park size, 111, 115, 119, 136, 138, 151
Philadelphia, 3, 8, 13, 14, 35, 163–166
Phoenix, 10, 160, 164, 165
Pittsburgh, 8
PlaNYC, 35
Portland, 8, 10, 13, 32
Property tax
  caps, 159, 160, 164, 167–173, 176, 179, 180, 195, 196
  laws, 158, 160, 166
Proposition 13, 21, 163
Public policies, 12, 33, 56, 76

Q
QCA, *see* Qualitative Comparative Analysis (QCA)
Qualitative Comparative Analysis (QCA), 2, 51–54, 109–113, 117, 118, 122, 131, 194

R
Race, 13, 14, 22–25, 27, 28, 36, 45, 78, 80, 90, 94, 114, 115, 139–141, 169
Rail Park, 166
Real estate market, 37, 38, 76, 78, 84, 134, 152, 167
Recipes, 54, 77, 96, 136, 139, 142, 193
Refurbishment, 28
Reinvestment, 2, 28, 75

Rent controls, 158, 159, 161–164, 166, 169, 170, 193
Renters, 14, 20, 21, 28, 63, 70, 83, 132, 149, 158–171, 174–176, 178, 179
Rent gaps, 21, 62, 187
Residential preferences
  density, 76, 79, 87, 88, 97
  segregation, 88, 92, 97
Richmond, 8, 12, 13
Risks, 24, 30, 31, 35, 47, 48, 50, 57, 65, 68–71, 83, 132, 143, 150, 151, 158, 159, 163, 165–167, 180, 184, 186–189, 191–195
Rose Kennedy Greenway, 166

S
San Diego, 3, 8, 35
San Francisco, 10, 25, 133
Scale, 32, 35, 36, 65, 75, 77, 137
Seattle, 8, 163–165, 171, 172
Sense of place, 115
Seoul, 35
The 606, 114, 119, 166
Social capital, 11, 20, 21, 114–115, 132, 186
Social efficiency, 19
Social mix, 11, 23, 186
Split tax rates, 34
Stockton, 95
Superfund, 12, 30, 31
Supply, 10, 15, 62, 66, 67

T
Tel Aviv, 20
Tenant protection, 158, 166
Time horizons, 5, 65, 69
Tipping point(s), 37
Toxics Release Inventory (TRI), 24
TPL, *see* Trust for Public Land (TPL), boundaries
TRI, *see* Toxics Release Inventory (TRI)
Trust for Public Land (TPL)
  boundaries, 115

U
UCC, *see* United Church of Christ (UCC)
United Church of Christ (UCC), 23
UPROSE, 30
Urban colonizer, 9
Urban greening, 32, 33, 105, 106, 110, 115, 120–122
Urban heat island (UHI), 6, 25
Urban pioneers, 9
Urban-planning, 33, 56

V
Vancouver, 35, 107

W
Washington, DC, 3, 8, 12, 13, 21, 160–165
Waukegan, 30
White, 3, 13, 14, 25, 27, 28, 74, 80, 153
White-non-Hispanic, 7, 9, 115, 126, 153
Willingness to pay (WTP), 30, 37
WTP, *see* Willingness to Pay (WTP)

Z
Zoning, 12, 35, 37, 38, 49, 55, 96, 137, 192

The manufacturer's authorised representative in the EU is Springer Nature Customer Service Centre GmbH, Europaplatz 3, 69115 Heidelberg, Germany. If you have any concerns regarding our products, please contact ProductSafety@springernature.com

Printed and bound by CPI Group (UK) Ltd, Croydon, CR0 4YY
30/09/2026
02225586-0001